T0320600

LISP LORE: A GUIDE TO PROGRAMMING THE LISP MACHINE

SECOND EDITION

LISP LORE: A GUIDE TO PROGRAMMING THE LISP MACHINE

SECOND EDITION

by

Hank Bromley
AT&T Bell Laboratories

and

Richard Lamson
Symbolics, Inc.

KLUWER ACADEMIC PUBLISHERS
Boston/Dordrecht/Lancaster

Distributors for North America:
Kluwer Academic Publishers
101 Philip Drive
Assinippi Park
Norwell, Massachusetts 02061, USA

Distributors for the UK and Ireland:
Kluwer Academic Publishers
MTP Press Limited
Falcon House, Queen Square
Lancaster, LA1 1RN, UNITED KINGDOM

Distributors for all other countries:
Kluwer Academic Publishers Group
Distribution Centre
Post Office Box 322
3300 AH Dordrecht, THE NETHERLANDS

Consulting Editor: Tom M. Mitchell

Library of Congress Cataloging-in-Publication Data

Bromley, Hank.
  Lisp lore.

  Includes index.
  1. LISP (Computer program language)   I. Lamson,
Richard. II. Title.
QA76.73.L23B75   1987   005.13 '3   87-3639
ISBN 0-89838-228-9

The Lexical Scoping example on page 52 is quoted from *Symbolics Common Lisp: Language Concepts*, Copyright © 1986 by Symbolics, Inc. Reprinted by permission.

Definitions from *The Hacker's Dictionary* Copyright © 1983 by Guy L. Steele. Reprinted by permission.

Text masters produced on Symbolics 3600^{TM}-family computers and printed on Symbolics LGP2 Laser Graphics Printers.

Printed in the United States of America.

Table of Contents

List of Figures

Preface to the First Edition

This book had its genesis in the following piece of computer mail:

```
From allegra!joan-b  Tue Dec 18 09:15:54 1984
To: sola!hjb
Subject: lispm

Hank, I've been talking with Mark Plotnik and Bill Gale
about asking you to conduct a basic course on using the
lisp machine.  Mark, for instance, would really like to
cover basics like the flavor system, etc., so he could
start doing his own programming without a lot of trial
and error, and Bill and I would be interested in this,
too.  I'm quite sure that Mark Jones, Bruce, Eric and
Van would also be really interested.  Would you like to
do it?  Bill has let me know that if you'd care to set
something up, he's free to meet with us anytime this
week or next (although I'll only be here on Wed. next
week) so we can come up with a plan.  What do you think?

Joan.
```

(All the people and computers mentioned above work at AT&T Bell Laboratories, in Murray Hill, New Jersey.) I agreed, with some trepidation, to try teaching such a course. It wasn't clear how I was going to explain the Lisp Machine environment to a few dozen beginners when at the time I felt I was scarcely able to keep *myself* afloat. Particularly since many of the "beginners" had PhD's in computer science and a decade or two of programming experience. But the need was apparent, and it sounded like fun to try, so we had a few planning sessions and began class the next month.

From early January through late March we met once a week, about a dozen times in all, generally choosing the topic for each session at the conclusion of the previous one. I spent the last few days before each meeting throwing together lecture notes and a problem set (typically finishing shortly *after* the announced class time). By the end of the course, the students had attained varying levels of expertise. In all likelihood, the person who learned the most was the instructor; nothing provides motivation to figure something out like having committed oneself to talking about it.

After it was over, another co-worker saw the sizable pile of handouts I had generated and proposed that it would make a good book. He offered to contact a publisher he had recently dealt with. I was at first skeptical that the informal notes I had hurriedly concocted would interest a reputable academic publisher, but after taking another look at the materials that had sprouted, and discussing the matter, we agreed that quite a few people would find them valuable. I've spent the last few months filling out and cleaning up the pile, and Presto, change-o. My "set of handouts" is "a book."

There are a number of people who have, in one way or another, consciously or otherwise, helped create this book. Ken Church was instrumental in arranging my first experience using the Lisp Machine, and later was responsible for bringing me to Bell Labs. He also taught a course here, before I came, which laid some of the groundwork for my own course. Eva Ejerhed, in a rare act of faith, hired me to work on a Lisp Machine thousands of miles from the nearest expert assistance, without my having ever touched one. Joan Bachenko and Bill Gale first suggested I teach a course at the Labs. Many of my colleagues who served as experimental subjects by participating in one of the three trials of the course provided useful comments on the class handouts; among those whose contributions I particularly recall are Mark Liberman, Jeff Gelbard and Doug Stumberger. Ted Kowalski first broached the idea of making a book from the handouts, and also – with Sharon Murrel – supplied lots of assistance with the use of their *Monk* text formatting system. Wayne Wolf suggested improvements to my coverage of managing multiple processes. Jon Balgley, of Symbolics, Inc.,[1] wrote a helpful review of one version of the manuscript. Valerie Barr introduced herself to the Lisp Machine by actually working through an entire draft, making a great many valuable observations along the way. Mitch Marcus and Osamu Fujimura, my supervision at the Labs, were most understanding about the amount of time I put into this project. Carl Harris was an obliging and patient Publisher. Finally, Symbolics, Inc. graciously allowed me to quote extensively from their copyrighted materials, and Sheryl Avruch of Symbolics made possible the distribution of a tape to accompany this book.

---

[1] Symbolics, Symbolics 3600, Symbolics 3640, Symbolics 3670, and Document Examiner are trademarks of Symbolics, Inc. Zetalisp(r) is a registered trademark of Symbolics, Inc.

I would like to hear about any problems readers have while working their way through the text. Please don't hesitate to mail me any of your comments or suggestions.

Hank Bromley

December, 1985

computer mail: US mail:[2]

hjb@mit-mc (arpa)   AT&T Bell Laboratories
alice   room 2D-410
research } !sola!hjb (uucp)   600 Mountain Avenue
allegra   Murray Hill, NJ 07974

---

[2]As of September, 1986, Hank is no longer working for AT&T. His new address is:

Hank Bromley
Martha's Coop
225 Lake Lawn Place
Madison, WI 53703

Preface to the Second Edition

I received my copy of *Lisp Lore* back in July directly from Hank; we had met at a course taught by Symbolics in Cambridge and he had mentioned it to me. Immediately, I recognized its value. Unfortunately, much of it was soon to be made obsolete by the issuance of Release 7.0, which was scheduled for a little over two months after its publication. I wished I had had time to review it before publication.

Two days later, I received a copy of the following piece of computer mail:

```
From: hjb.sola%btl.csnet@CSNET-RELAY.ARPA
Date: Mon 21 Jul EDT 1986 18:51
To: SLUG@R20.UTEXAS.EDU
Subject: masochist, I mean writer, needed

The publisher of my book ("Lisp Lore:  A Guide to Programming
the Lisp Machine") would like to do a revised-for-Release-7
version.  I don't have the time to do the revision.  If you
or anyone you know might be interested, have them call me
(201/582-4377), or send me mail, or call Carl Harris at
Kluwer Academic Publishers (617/871-6300).
```

Well, here it is, months later, and I'm getting my "wish." I hope this edition is as valuable as I found the first. I've certainly had fun writing it.

An enormous number of people have contributed to my ability to get this work done. First, of course, Hank Bromley, who turned his manuscript over to me, both emotionally and electronically. A number of my colleagues at Symbolics have read the various drafts and commented quite helpfully: Muffy Barkocy (who also drew the card font for the solitaire program), Lois Wolf, Carmen Silva, Debbie Ward, Robert ("BigBob") Westcott, Jon Balgley and Allan Wechsler. Thom Whitaker greatly aided my efforts to make Scribe make the book printable. My managers during this project, Larry Rostetter and Jim O'Donnell, were extraordinarily supportive. And my family and loved ones, especially Joan Freedman, have given me encouragement and massages when all else failed.

I would certainly like to hear about problems you have while reading this book and its accompanying examples. Please don't hesitate to send me comments or suggestions. I imagine there will be later editions as this one becomes obsolete.

By the way, you might be slightly confused by the fact that both Hank and I wrote sections of this book in first person. I've adapted as much of Hank's text as possible, including all his "I"s. However, I did write the following chapters from scratch: 6, 9, 10, 11 and 13. Most of the other chapters have had a pretty substantial updating.

Richard Lamson

November, 1986

computer mail: US mail:

rsl@Symbolics.ARPA   Symbolics, Inc.
rsl@E.SCRC.Symbolics.COM   25 Van Ness Blvd.
                                                         San Francisco, Calif 94102

ACKNOWLEDGMENTS

The authors gratefully acknowledge AT&T Bell Labs and Symbolics, Inc., for their research environment support for such forward-looking and creative activities as produced this book.

We would also like to thank the companies listed below for their permission to use the following registered trademarks or symbols:

> **Symbolics, Symbolics 3600, Symbolics 3670®, Symbolics 3675, Symbolics 3640, Symbolics 3645, Symbolics 3610, Symbolics 3620, Symbolics 3650, Genera, Symbolics-Lisp®, Wheels, Symbolics Common Lisp, Zetalisp®, Dynamic Windows, Document Examiner, Showcase, SmartStore, SemantiCue, Frame-Up, Firewall, S-DYNAMICS®, S-GEOMETRY®, S-PAINT, S-RENDER®, MACSYMA, COMMON LISP MAC-SYMA, CL-MACSYMA, LISP MACHINE MACSYMA, MACSYMA Newsletter® and Your Next Step in Computing®** are trademarks of Symbolics, Inc.

> "Unilogic" and "Scribe" are registered trademarks of Unilogic, Ltd.

> **DECnet** is a trademark of Digital Equipment Corporation.

> **Multics** is a trademark of Honeywell, Inc.

> **UNIX** is a trademark of AT&T Bell Laboratories.

> **VAX** and **VMS** are trademarks of the Digital Equipment Corporation.

LISP LORE: A GUIDE TO PROGRAMMING THE LISP MACHINE

SECOND EDITION

1. Introduction

The full 13-volume set of documentation that comes with a Symbolics Lisp Machine is understandably intimidating to the novice. "Where do I start?" is an oft-heard question, and one without a good answer. The thirteen volumes provide an excellent reference medium, but are largely lacking in tutorial material suitable for a beginner. This book is intended to fill that gap. No claim is made for completeness of coverage – the thirteen volumes fulfill that need. My goal is rather to present a readily grasped introduction to several representative areas of interest, including enough information to show how easy it is to build useful programs on the Lisp Machine. At the end of this course, the student should have a clear enough picture of what facilities exist on the machine to make effective use of the complete documentation, instead of being overwhelmed by it.

The desire to cover a broad range of topics, coupled with the necessity of limiting the amount of text, caused many items to be mentioned or referred to with little or no explanation. It's always appropriate to look up in the full documentation anything that's confusing. The manuals are perfectly adequate reference materials, as long as you know what you want to look up. The point in *this* text is rarely to explain what some specific function does in isolation – that's what the manuals are good for. The focus here is on how to integrate the isolated pieces into real applications, how to find your way around the landscape, how to *use* the multitudinous features described in such splendid detail in the 13 volumes. The manuals provide a wonderfully thorough, but static, view of what's in the Lisp Machine environment; I've tried to provide a dynamic view of what that environment looks like in action, or rather in inter-action with a human.

The book assumes some background in Lisp; the reader is expected to have experience with some dialect of the language. If you lack such experience, you may want to do a bit of preparatory study.[1] This book concentrates on those aspects of the Lisp Machine language *(Symbolics Common Lisp)* which are not found in most dialects, and on the unique overall program-ming environment (called *Genera*) offered by the Lisp Machine. No experience with the Lisp Machine itself is assumed.

Finding an ideal order of presentation for the various topics would be difficult. Many topics are interdependent, such that knowing either would help in figuring out the other. Present-ing them simultaneously would only confuse matters, so I've had to settle on one particular linear sequence of topics. It may seem natural to some readers and bizarre to others. I've tried to identify places where it might be helpful to look ahead at sections further on in the text, but I'm sure I haven't found them all, so don't hesitate to engage in a little creative re-ordering if you feel the urge. One chapter whose position is problematic is that on Flavors. A great deal of the Lisp

---

[1] Two widely available sources you may find well worth your time are *Lisp* (2nd edition), Winston and Horn, Addison-Wesley, 1984, and *Structure and Interpretation of Computer Programs*, Abelson and Sussman, MIT Press, 1984.

Machine environment depends on Flavors, including the window system and the process scheduler. I have chosen to place it after the chapter that introduces both of those, because you don't really need to know flavors to get the concepts, and it really belonged after *Flow of Control*. Feel free to sneak looks at it at any time, though. Also, I've sprinkled references to the Symbolics documentation liberally throughout the book; feel free to chase down anything that interests you.

I've adopted a rather informal tone for most of the text: people learn better if they're relaxed. Just let me caution you that "informal" doesn't mean "sloppy." There are few extra words. Lots of information is present in only one place, and apparent only if you read carefully. If you get fooled by the informality into thinking you can scan half-attentively, you'll miss things.

It must be emphasized that learning to use the Lisp Machine is more a matter of learning a way of thinking than of learning a set of specific programming constructs. No amount of time spent studiously poring over documentation can yield the benefits gained from sitting at a console and exploring the environment directly. Time spent examining various parts of the system software with no particular goal in mind is anything but wasted. Once one has a feel for how things are done, an overview of how things fit together, the rest will follow easily enough. Most Lisp Machine wizards are self-taught; the integrated nature of the environment, and ready access to the system code, favors those who treat learning the machine as an interactive game to play.

With that in mind, a word or two of advice on the problem sets. Don't get too wrapped up in finding the "right answer." Many of the problems are, shall we say, "challenging;" they require knowledge not found in the text (and in some cases not even found in the manuals). You will need to investigate, often without knowing exactly what you're looking for. If the investigation fails to yield immediate results, I strongly recommend that rather than head straight for my solutions, you continue to investigate. Stick it out for a while, even if you don't seem to be getting much closer. You can't learn to speak a foreign language by consulting a dictionary every time you need a word you don't know – forcing yourself to improvise from what you do know is the only way. Floundering is an unpleasant but absolutely necessary part of the process, arguably the only part during which you're really learning. Similarly, you can't become a Lisp wiz just by assiduously studying someone else's code. Although seeing how an experienced programmer handles a problem is certainly useful, it's no substitute for struggling through it yourself. The problem sets are largely a ruse to get you mucking around on the machine. I don't really care if you solve them, as long as you come up with some ideas and try them out with an open mind.

The examples in the text (barring typos) are known to work in Release 7.0 of the Symbolics Genera software for the 3600 family of machines. A machine-readable copy of all the examples is available on a cartridge tape; it also includes all problem solutions which are too long reasonably to be manually copied from the text.

To order a copy of the tape, write to the following address (you may wish to use the order form at the back of this book) and include a check[2] for $40 made out to Symbolics, Inc. Instructions for loading the tape will accompany it.

---

[2]Unfortunately, Symbolics cannot accept purchase orders for this. Residents of the followings states please add the appropriate sales tax: AZ, CA, CO, CT, FL, GA, IL, KS, MA, MN, NJ, NM, NY, OH, PA, TX, VA, WA.

Software Release
Symbolics, Inc.
Eleven Cambridge Center
Cambridge, MA 02142

2. Getting Started on the Lisp Machine

2.1 Why Use a Lisp Machine?

Why would anyone use a Lisp Machine? It is expensive and difficult to learn, true, but there must be a reason, since so many of them are being sold.

The answer is that the Lisp Machine is a powerful computing environment, well designed for a wide variety of tasks. These include:

- *Programming*: The Lisp Machine has been designed with programmers in mind. The system development cycle of editing, compiling, testing and packaging for delivery is heavily optimized. Many features have been added during its decade of existence to make life blissful for programmers. These environment features lead to reduced development time for software.

- *Programming Experiments*: Many computer programs today are written before the problems they attempt to solve are completely understood. For example, consider the problem of making a computer understand human speech. It is not known exactly how humans accomplish this task. It is only by writing experimental computer programs that we can learn which methods work and which do not. The Lisp Machine offers an environment wherein experimentation is efficient and painless.

- *User interface experiments*: Even if the underlying model for how to solve the problem is well-understood, new methods of presenting information to and accepting input from people are often controversial. The Lisp Machine offers a toolkit of user interface functions which may be combined in new and exciting ways.

- *Delivery of systems*: While it is true that much of the software developed on the Lisp Machine may be moved to other computer systems, some of it remains dependent on the software environment. If a decision is made to deliver software on the Lisp Machine, more powerful tools, such as the user interface toolkit, may be applied to the solution of problems, simplifying maintenance and shortening delivery times.

In short, use of the Lisp Machine simplifies design and speeds coding and correction, and facilitates experiments in both user interfaces and programming methods.

2.1.1 Why This Book?

A powerful system is powerful at least in part because it provides experienced users with a lot of flexibility. This is certainly true of the Lisp Machine. As a result of great flexibility, though, the learning process is difficult, because of two obstacles:

- The environment is *intimidating*. There are an enormous number of commands, and the documentation set (all 13 volumes of it, and growing at each release) is huge.

- Part of the intimidation is *historical*. Many of the interesting things people wanted to do were hard to figure out, especially from the documentation. There were very few published examples, and few of them were well-written or easy to understand.

This book is an attempt to reduce the intimidation level. Part of that process is teaching a basic level of expertise, which will make many of your tasks easier. However, I also hope to teach you something more fundamental, namely how to learn more on your own.

I hope to teach you how to become an expert in using the Lisp Machine. When we're done, I hope:

- You will know how to write programs for the Lisp Machine, including large ones.
- You will be able to learn new things about the system.
- You will understand a lot of features which are intimidating but not very hard to use.

2.1.2 Looking Ahead

The rest of this chapter is an introduction to the Lisp Machine environment. It is best to read it in front of the console. Don't be afraid to try things out; it's very difficult to make irrecoverable mistakes.

- The section *The Keyboard* gives a guided tour to typing on the Lisp Machine's console.

- *Typing to a Lisp Listener* describes one of the ways you get the machine to do what you want.

- In *Getting Around the Environment,* you will learn how to get to other facilities the Lisp Machine environment offers.

- The section *The Mouse* describes the use the other input device attached to your console.

- In *The System Menu* I outline another way to get around the environment.

- In *The Monitor,* the mysteries of the screen display will be discussed.

- *The Editor* provides a first lesson in using the Lisp Machine editor *(Zmacs).*

- In *The Compiler and the Debugger,* I will introduce testing and modifying your programs.

- In order to use your machine at all, it must be running and you must be logged in. In *Getting Started,* I'll talk about what to do if your machine isn't already running, or doesn't know who you are.

- *A Word About Work Style* contains a few reassuring remarks.

- In *This and That* will be everything else I wanted to tell you that didn't appear elsewhere in this chapter.

A final word: it's easier than it looks. In a very short time you will start doing useful work, using the Lisp Machine effectively.

2.2 The Keyboard

Sit down at the console of your machine. You will see many keys which don't appear on a standard keyboard. Much of what you need to know to start using a Lisp Machine boils down to knowing what the various funny keys do.[1]

Apart from the keys for the standard printing characters (white labels on grey keys), the keyboard has two kinds of special keys. The beige keys with grey labels (shift, control, meta, super, hyper, and symbol) are all used like the "shift" key on a normal typewriter – you hold them down while striking some other key. These *modifier* keys may be used singly or in combination. So "control-meta-K" means type K while holding down control and meta. You may use either of the two modifier keys with the same name, just like the "shift" key on regular typewriters.

Symbolics uses a standard set of abbreviations for the various modifier keys. They're all just what you'd expect except that the abbreviation *s* stands for *super* rather than *shift*. *Shift* is abbreviated *sh*.

The beige keys with white labels are special function keys, and are typed like standard printing characters. Some of them stand alone and have obvious meanings, like Clear-Input and Help. These keys, of course, can be modified with any of the modifier keys; some programs, for example, do something special when they read meta-Help or control-Clear-Input. I recommend that you use the Help key a lot. The information supplied depends on the context, but typing Help usually tells you what sort of input is wanted by the program you're typing to.

---

[1] In order for much of this discussion to "work," the machine must be running. See the section "Bringing the Machine up," page 25.

Some of these special keys are used as prefix characters, such as Select and Function. That is, "Select E" means to strike Select, and *then* strike E. And "Select c-L" means to strike Select, and then hold down control and strike L.

The Select key is used to select the program you wish to use. It allows access to such programs as the Editor, the mail reading/sending program, the remote login facility, and others. Which program it selects depends on the next character you type. For further information, try pressing Select Help. See the section "Getting Around the Environment," page 15.

The Function key, like Select, dispatches off the following keystroke. RFunction Help displays a list of the options. The most commonly used are Function F ("finger"), to find out who's logged in on the various machines, Function H ("host status"), for a quick look at the status of all the hosts on the local Chaosnet, and Function S to select a different window. The exact behavior of many of the Function options is controlled by an optional numeric argument; you pass the argument by pressing one of the number keys after the Function key and before the chosen letter, *e.g.*, Function 0 S.

The Suspend key generally causes the process you are typing to to enter a "break loop", that is, the state of its computation is suspended and a fresh read-eval-print loop is pushed on top of the current control stack. The Resume key will continue the interrupted computation. Suspend takes effect when it is read, not when it is typed. If the program isn't bothering to check for keyboard input, pressing Suspend will do nothing (until it *is* read).

c-Suspend does the same thing as Suspend, but always takes effect immediately, regardless of whether the program is looking for keyboard input.

m-Suspend, when read, forces the reading process into the

debugger. The debugger is described later. (See the section "The Compiler and the Debugger," page 23.) When you're done looking around you can continue the interrupted computation with Resume.

c-m-Suspend is a combination of c-Suspend and m-Suspend. It immediately forces the current process into the debugger.

The Abort key is used to tell a program to stop what it's doing. The exact behavior depends on what program you're typing to. A Lisp Listener, for instance, will respond by throwing back to the nearest read-eval-print loop (the top level or an intervening break loop). Like Suspend, Abort only takes effect when read. If the program isn't waiting for keyboard input, you need to use c-Abort instead.

m-Abort, when read, throws out of all levels and restarts the top level of the process. c-m-Abort has this effect immediately.

2.3 Typing to a Lisp Listener

A *Lisp Listener* is a window with a Lisp interpreter running in it. It reads a Lisp expression (or *form*) from the keyboard, evaluates it, prints the returned value(s), and waits for another expression. Booting a machine leaves you in a Lisp Listener. Whenever you're not in a Lisp Listener you can get to one by typing Select L.

While waiting for input, Lisp Listeners usually display "Command:" as a prompt. The presence of this prompt indicates that the *Command Processor* (CP) is active; it provides a convenient interface to many frequently called Lisp functions. (The name of a CP command won't necessarily be the same as the name of the corresponding Lisp function.) CP commands don't use the same parentheses syntax as Lisp expressions do.

You simply type the name of the command (one or more words) followed by any arguments to the command, and finish with the Return key. But you needn't type the name of the command in its entirety – all that's required is enough to uniquely identify which command you mean.

The CP command Help (*i.e.*, type out the letters h, e, l, p, and hit Return) lists all the defined commands. Once you have started typing a command, pressing the Help *key* while partway through a command will display a list of only those commands which match your input thus far.

Volume 1 of the documentation describes all the CP commands present in the software distributed by Symbolics (See the section "Dictionary of Command Processor Commands" in *User's Guide to Symbolics Computers*.). You can define more of your own. One command which may be particularly valuable to new users is Show Documentation. You specify some topic you want looked up in the manuals and it displays a facsimile of that portion of the documentation on your screen.

You may be wondering how the Command Processor knows whether you intend your typein to be interpreted as a CP command or as a Lisp expression. In its normal mode, the CP looks at the first character on the line, and if it's a letter, the CP tries to interpret the line as a command. If this succeeds, the line is a command.

If there is no command by that name, the CP next looks to see if you have typed the name of a symbol with a value. If there is one, the input is interpreted as a Lisp expression to be evaluated, or *form*.

There are occasions when you want to type a form that begins with a letter, but is not a symbol.[2] The solution here is to type a comma at the beginning of the line. The comma has special meaning for the command processor: it forces whatever follows to be interpreted as a Lisp expression, regardless of what the initial character is.

If you'd like to know about some other features that are available whenever you're typing to a Lisp Listener and you don't already feel as though you've seen more than you can possibly remember, you might look ahead a few chapters: See the section "The Input Editor and Histories," page 83. It'll make life much easier as you take on the first few problem sets.

2.4 Getting Around the Environment

The Lisp Machine environment actually consists of a number of environments, called *activities*. Each activity is associated with a particular window and its own way of interpreting what you type.

For example, the Lisp Listener activity has a Lisp Listener window, and interprets what you type as being either commands or Lisp forms. The Zmacs activity, on the other hand, displays an editor window, and interprets your input as editor commands.

There are several ways to select an activity. One of them is the Select Activity CP command. If you type

---

[2]For advanced readers: this happens, for example, when you want a package prefix to apply to an entire form, as in tv:(make-window 'window). If you don't understand this, you don't need to worry about it.

Select Activity Lisp

you will wind up looking at your Lisp Listener window. As usual, pressing the Help key after typing the command name will tell you your choices.

One of the quickest ways to switch among activities is the Select key. To find out what your options are, type Select Help. The display shows you that, among other programs that may be reached in this way, you can get a Lisp Listener by typing Select L, and an editor by typing Select E. This list is by no means fixed. Users may add their own programs to the list quite easily. Here are brief descriptions of the programs that are already in the select list on a freshly booted Lisp Machine:

C Converse   a convenient way to send and receive messages from users currently logged-in on other machines (Lisp or otherwise)

D Document Examiner
                  a utility for finding and reading online documentation; everything in the 13-volume manual is available here

E Editor   the powerful Zmacs editor, like Emacs plus much more

F File System Editor
                  various display and maintenance operations on the file system of the Lisp machine or of other machines

I Inspector   the structure editor for displaying and modifying Lisp objects

L Lisp   a Lisp Listener

M Zmail   a comprehensive mail-reading and sending program.

N Notifications a display of all "notifications" (messages from running programs) you've received

P Peek   a status display of various aspects of the Lisp Machine

Q Frame-up   a utility program for laying out the display for user-written programs.

T Terminal   a program which allows you to use the Lisp Machine as a terminal to log in to other computers

X Flavor Examiner
                a convenient way to find out about different flavors (active objects), their message-handlers, and their state variables.

The Function key can also be used to select among activities, by choosing which window is selected. Try pressing Function Help to see what Function S and Function B do.

Finally, the mouse may also be used to select activities. See the section "The System Menu," page 18.

2.5 The Mouse

The mouse is used for pointing to objects displayed on the screen. You "point" by moving the mouse on the table while watching its cursor move on the screen. As the mouse cursor moves, certain objects on the screen will be highlighted, by drawing a little box around them; these objects are said to be *mouse sensitive*. Just pointing at objects is fun, but doesn't say

much about what you want to do with (or to) them. To communicate your intent to the system, you can press any one of
its buttons, either quickly releasing it (called *clicking*) or holding the button down. Most of the usual uses of the mouse involve clicking.

As of Genera 7.0, most items displayed on the screen are mouse
sensitive, which means that clicking on them will do something
of interest. What clicking will do depends on where you click,
which keyboard modifier keys you happen to be pressing, and
what your current activity is doing at the moment.

How can you tell what will happen when you click? The little
white-on-black area near the bottom of the screen documents
what the mouse will do. For example, as I type this into
Zmacs, the mouse documentation lines say:

```
Mouse-L: Move point; Mouse-M: Mark word; Mouse-R: Editor ...
To see other commands, press Shift, Control, Meta-Shift, ...
```

If I press Shift, the display changes as follows:

```
sh-Mouse-L: Move to Point; sh-Mouse-M: Save/Kill/Yank; ...
To see other commands, press Shift, Control, Meta-Shift, ...
```

While some of these may seem cryptic to you at first, you will
quickly learn to find the ones of interest.

2.5.1 The System Menu

The system menu is a collection of operations that people find
useful. You can cause the system menu to appear at any time

by clicking Shift-Mouse-Right.[3]

The system menu contains three columns of options which may be selected. The first column holds window-system operations, such as creating a new window, selecting a window, or editing the position of the windows on the screen. The second column contains commands pertaining to the current window and the process which owns it. These commands include changing the shape of the window, hardcopying it, killing it, or resetting or halting (*"arresting"*) its process. The third column contains the names of activities which may be selected by using the system menu.

2.6 The Monitor

In addition to the mouse documentation, the bottom of the screen contains a lot of other useful information. From left to right, we have:

- the date and time

- the user-id of the currently logged-in user, if any

- the current *package*[4]

---

[3]In general, a mouse click which is performed by holding down the *Shift* button is equivalent to clicking twice in rapid succession. In previous releases of the system, in fact, people were told to "double-click Right" to get to the system menu. This mode of using the mouse is considered obsolete, however, and might be removed from the system in a future release.

[4]The set of all symbols is partitioned into packages, to minimize name conflicts – a cold-booted machine starts out in the "user" package, which is where you'll probably do most of your work at the beginning.

- the state of the current process ("User Input" means awaiting keyboard input)

All the way on the right is a special little display area which shows you other activities which might be happening on your machine. At different times, this area might contain the names of any files that are open for reading or writing, other "progress notes" about things which are happening on your machine, or a notice of what network services have been invoked on your machine by some other machine.

Underneath the line of text you can sometimes see a few thin horizontal lines. These are the *run bars*. The one immediately under the process state appears when some process is actively running. The one a bit to the left of that, midway between the process state and the current package, indicates that you are paging, waiting for something to be brought in from disk. You will see other run bars less often. They are related to garbage collection, and to saving a snapshot of the Lisp environment on the disk.

One last bit of information on the monitor. To adjust the brightness, hold down the Local key and press B for brighter or D for dimmer. [Yes, the key has the wrong color lettering; it's really a shift key and not a prefix. Sorry.] The volume of the *beep* may be adjusted by holding down Local and pressing "L" for louder or "Q" for quieter.[5]

---

[5]Both of these may also be adjusted by using the special form **setf** on the functions **sys:console-brightness** and **sys:console-volume**, respectively.

2.7 The Editor

The Lisp Machine's Zmacs editor is based on Emacs, a text editor written at MIT. You use the editor for typing programs and other text files.

For typing programs, you *could* use the Lisp Listener, but the editor is easier to use, and more powerful. When editing a Lisp program, for example, the editor helps you in balance your parentheses and indent your code so it's easier to read. You can compile prototype or experimental code directly in the editor, and immediately test it. When the compiler warns you about source errors, the editor can be used to examine the error messages while fixing the source (using the Edit Compiler Warnings editor command). And, of course, once you are done modifying the source to your software, you can use the editor to save it on disk.

The editor is also good for reading programs that other people have written, including the source for the system as shipped by Symbolics. The editor command m-. (pronounced "meta-dot," or [infrequently] "meta-period") prompts for the name of a function, variable or flavor; you may either type in the name, or click on it with the mouse. When the command completes, you will be looking at the source of the definition in question. Similarly, there are editor commands for listing and editing the callers of various functions and flavor components.

Some other things you might use the editor for include:

- Typing documentation or other text.
- Editing file directories.
- Sending mail.

As is true of most of the rest of the system, the editor is self-documenting. The built-in editor commands are bewildering in number, and the total number of available commands is continually growing because it's fairly easy and very tempting to add new ones. The Appendix lists the most basic commands, but by far the best way to find out what's around is to get used to using the online documentation. Some aspects of the Lisp Machine can be mastered by reading the manuals, but the editor is not one of them. Press the Help key to an editor window, and press Help again. Start exploring. The most commonly helpful of the help options are A (Apropos), C (Command), and D (Describe). To get started, use Help C on c-X c-F and on c-X c-S. You should also try Help A on "compile."

If you type Suspend, you get a Lisp Listener which starts at the top of the screen and grows as you need it. This funny window is called the *typeout window*.[6] Resume returns to the editor.

Dired is a utility for editing directories. It is invoked by typing m-X Edit Directory or c-X D. Call it on some directory and press Help. (Keep in mind that if it's a Lisp Machine directory, there might be no security to keep you from deleting absolutely anything. By default, the Lisp Machine file access control mechanism is turned off. See the section "Access Control Lists" in *Reference Guide to Streams, Files, and I/O*.)

---

[6]One problem with the editor typeout window is that when you're using it to debug a program which is in the editor buffer, you can't use the editor to read or modify it. The debugger provides a way to edit the source of a function which is not behaving properly, which is harder to use from the editor's typeout window because it throws away your debugging state. See the section "The Compiler and the Debugger," page 23.

2.8 The Compiler and the Debugger

The compiler transforms Lisp source into a binary form which executes rapidly on the Lisp Machine. While it is true that you can execute your Lisp code directly (in "interpreted" form), the compiler is very fast, and produces much faster code; also, the debugger has been optimized for use with compiled code.

The compiler operates in two modes: "core" and file. In "core" mode, the binary version of your source is placed directly into "core" (*i.e.*, directly into your Lisp environment). In file mode, the binary version is put into a file, which may be loaded immediately into your environment or later into some other environment (perhaps on a different Lisp Machine). Most frequently, you will compile to "core" by using the editor command c-sh-C. To compile a source file into a binary file, you can use the Compile File (m-X) command in the editor or in the CP. To load a compiled file, use the Load File command in the editor or the CP.

Often your program will not work properly the first time. Since the Lisp Machine checks for many errors implictly,[7] you will probably wind up in the debugger fairly often while writing and testing your software. You will not usually wind up in the debugger just from typing something the Lisp Machine doesn't "understand;" usually, you get an informative diagnostic error and are either permitted to correct your mistake or to try a different command. Don't be afraid of the debugger, though. The debugger is your friend.

The debugger prompt is a small right-pointing arrow. Once

---

[7] such as using data of the wrong type for most built-in operations, undefined variables, *etc.,*

you have that, all kinds of commands are available for moving up and down the stack,[8] getting information about the different frames on the stack, restarting execution with or without modifying arguments and variable values, *etc.* Try the Help key and see what you can find out. Besides all the special commands, any normal text you type will be evaluated by the Lisp interpreter.

One debugger command you will find extremely useful is control-E, which takes you to the editor and shows you the source of the function you are debugging. This command works even while using the editor's typeout window, but will throw away the state of the debugger and the error you are attempting to understand when you use it in the typeout window. Thus, while it is often useful to try out your code in the typeout window, you must keep this restriction in mind when you have complex errors to puzzle out.

2.9 Getting Started

It should be obvious that before you can use your machine, it must be up and running. In addition, before you can use many of the interesting facilities of the Lisp Machine, you must be logged in.

---

[8]When a Lisp function calls another, the Lisp Machine must remember the calling function, its local variables, *etc.* It does this in a data structure called the *stack*, sometimes called a "push-down stack." The computer literature apocryhpa claim it is so-called because it resembles stacks of dishes in a cafeteria; when you add more, the others are pushed down. You might even see a stack called a "push-down list," or PDL, in rare places in the system software.

2.9.1 Bringing the Machine up

First of all, you need to be able to tell if the machine is up. If the screen is dark, make sure the console is turned on, and then hold down the Local and B keys until it gets bright enough to read.[9]

Now, look at the display. When the machine is running, the bottom left-hand corner has the time of day, and is updated once per second. If this doesn't appear, or isn't being updated, the machine is probably not running. If the screen is covered with "snow," the machine's power is probably turned off.

When the Lisp Machine isn't running, your typing is read by another processor within the machine, called the *Front End Processor*, or FEP. This is a microprocesor whose job it is to bring up the machine, and to perform certain helping tasks while the Lisp Machine is up (like listening to the keyboard and running the cartridge tape drive). When the FEP is waiting for you to type a command, it prompts you with the string "FEP Command:".

Immediately after you turn on the power to your machine, the FEP initializes itself, and suggests that you type Hello. When told to so do, type Hello and a carriage return.

To bring up the machine from the FEP, you usually use the Boot command. Type Boot and carriage return, and the machine will read the commands in a file named "Boot.boot," which should actually bring up the machine. For further information: See the section "Cold Booting" in *Site Operations*.

---

[9]In Genera 7.0, the system automatically dims the screen if nobody has touched the keyboard or mouse for a few minutes. Try pressing the shift key or jiggling the mouse to see if the screen brightens.

Cold booting a machine which has been in use wipes out all the changes that have been made to the world, including, for example, your editor buffers and so forth. Once in a while, you have a hardware or software problem which halts your machine while you're using it. If you cold boot, any work you have done which has not been written on disk is lost.

Fortunately, there are a couple of other choices. If you type Continue to the FEP, the machine will be restarted where it was. If your machine crashed because of some serious system error, this isn't likely to work – you will probably be thrown right back into the FEP. Once in a while this works; it's always worth trying, because there's no harm in it.

Another option is called *warm booting*. The Start command, by itself, attempts to warm boot the machine. It tries to restart all of the machines active processes while preserving the state of the Lisp environment (*i.e.*, function and variable bindings). This is something of a kludge.[10] It can put things into an inconsistent state, and is something of a last resort, but it is

---

[10]**KLUGE, KLUDGE** (*klooj*) *noun.*

1. A Rube Goldberg device in hardware or software.

2. A clever programming trick intended to solve a particularly nasty case in an efficient, if not clear, manner. Often used to repair BUGS. Often verges on being a CROCK.

3. Something that works for the wrong reason.

4. *verb*. To insert a kluge into a program. "I've kluged this routine to get around that weird bug, but there's probably a better way." Also "kluge up."

5. A FEATURE that is implemented in a RUDE manner.

(*The Hacker's Dictionary*, Guy L. Steele, Jr.,*et al*, Harper & Row, Publishers, New York, 1983.)

sometimes the only way to get a wedged machine started again, short of wiping the environment clean, and losing whatever work was in progress. Often, a good thing to do when you warm boot is to use the Logout command, which asks if you want to save your file buffers, and then halt the machine (using the Halt Machine command) and cold boot.

2.9.2 Logging In

Logging in is extremely simple. Use the Login command. For example, this is what the screen looks like when I log in:

```
Command: Login rsl
```

It's important to keep in mind the difference between a local login to the Lisp Machine and remote logins to other machines being used as file servers. Local logins are controlled by a database called the namespace. To login locally with a certain user-id requires that there be an entry in the namespace for that user-id. It usually does not require a password, as by default there is no internal security on the Lisp Machine.

Many things on the Lisp Machine can be done with no one logged in. Some operations, however, do require that someone be logged in. Modifying the namespace, for instance, is one of these operations. How, then, you may ask, do you create a namespace entry for yourself if you can't modify the namespace unless you're logged in, and you can't log in unless you're in the namespace?

One option would be to log in as someone else so you can create a namespace entry for yourself, and then log in as yourself. In fact, there is a "user" in every namespace for that very purpose, named "Lisp-Machine."[11] Once you're logged in,

---

[11]You can log in as "Lisp-Machine" by saying **(si:login-to-sys-host)**.

you can use the `Edit Namespace Object` comand to create your
user object.

However, nothing so underhanded is really necessary. The
`Login` command, if you attempt to log in as someone who is not
in the namespace, gives you the option of creating a namespace
entry for that user-id:[12]

```
Command: Login george
The user named "george" was not found:
Do you want to log in as a-new-user on some specific host?
    (Y, N or R) Yes
Host to log in to: your-file-server
Do you wish to add george to the user database?
    (Y or N) Yes
```

You will enter the namespace editor, which will allow you to
fill in your user entry. For documentation on the namespace
editor: See the section "Updating the Namespace Database" in
Networks.

Whenever you log in to a Lisp Machine, unless you explicitly
specify otherwise, it tries to find your personal initialization file
and load it into the Lisp environment. This is a file containing
any number of Lisp forms which customize the machine for
you. They will typically set some variable values, define functions and commands, and perhaps load other files. Where your
machine looks for your "init file" depends on what you
specified for your *home host* in your namespace entry. A file
named "lispm-init" in your home directory on that file server is
loaded using the usual rules: first it looks for a compiled file
(with type "bin"), and if that fails it looks for a source file
(with type "lisp").

---

[12]For this example, all <u>underlined</u> text is what the user types; everything not
underlined is typed out by the computer.

The issue of remote logins arises whenever you use the network to try to do something on another computer from the Lisp Machine, like read or write a file. If the remote host is a Lisp Machine, it won't ask for a password and your local machine can take care of establishing the connection with no intervention on your part.[13] If the remote host is the sort that believes in security, like Unix or VMS, it won't let your Lisp Machine do anything until you type an appropriate login id and password. Your local machine will pass the request right along to you. It's essentially a matter between you and the remote host – the local machine doesn't care what username you use on the remote machine, nor whether it's one that exists in the namespace. The local machine is just a messenger in this case. It will, however, try to be helpful. If you specify in your namespace entry what user-names you want to use on the various remote hosts, the local machine will try those first, even if those names are arbitrary nonsense as far as the local machine can tell. If this fails, you can always override the default usernames.

While we're on the subject of remote file systems, there's the question of whether you should keep your files on a Lisp Machine or some other sort of file server. It depends on what sort of setup you have – how much disk space in what places, how many users, *etc.* Wherever you store your files, make sure they are regularly backed up to tape. You can imagine horror stories of hardware failures which wipe out months or years of work.

---

[13]Unless someone has turned on the Access Control mechanism on the Lisp Machine. In such cases, the Lisp Machine will ask for passwords the way any other security-minded system does.

2.10 A Word About Work Style

OK, you're almost ready to have some fun with your Lisp Machine. But first, a final word.

There are an almost mind-numbing number of different ways to do anything you can think of. For example, some people really like being able to point and click with the mouse, while others don't like taking their hands off the keyboard. At every level of detail, from how your programs do their job to how you manage your file system space, there are choices to make.

Some examples:

- Text editing: The text editor provides a large number of ways, for example, to move the cursor. Some examples include moving forward/backward by single characters, lines, words, sentences or Lisp expressions. You can also click on a place in the visible text, using the mouse, and the cursor will move there.

- Debugging: There are two different user interfaces to the debugger, the keystroke debugger (described above) and the *window debugger*. Each one has its fans and detractors. The window debugger displays the current state in a convenient format, and allows exploration of the error via mouse clicks. The keystroke debugger is more esoteric, but can be faster once you learn your way around it.

- Directory editing: There are two independent programs for viewing and editing the contents of a file system directory. The File System Editor (Select-F) displays the contents of a directory; the user updates the display and performs operations on files using the mouse. The Directory Editor (DIRED) is part of the Zmacs editor, and its commands are primarily keyboard commands. It fits seamlessly into the normal text editor.

[In the last two examples, the keyboard option is somewhat harder to learn, but is much faster to use once you're good at it.]

By now, it should be clear that there is no one right way to use the system which works for everybody. Each user will develop a style which works for her or him. Don't be afraid to experiment.

2.11 This and That

I want to discuss a couple of subjects near and dear to my heart: bug reporting and file backup. It is essential that someone at your site be responsible for these two operations.

2.11.1 Problem Reporting

There should be someone at your site who is your contact with Symbolics. Perhaps you will be the person.

There are a number of useful facilities in the Lisp Machine environment for reporting problems. In the debugger, the command c-M will set you up to send mail concerning your problem. The mail will already contain the version of the system you are using, and a copy of the stack trace. You merely have to type in what you did to cause the problem and type End.

There are also Report Bug commands in the CP and the editor. They also set you up to send mail about your problem. In all these mail buffers, you can put a file pathname into the mail if you think your problem is related to its contents, using the Add File Reference mail command, or include the contents of the file using the Insert File command.

If your machine crashes, you can report information about the problem in your mail once you bring the machine back up. Use the Insert Crash Data mail command.

If you are the person who is your site's liaison with Symbolics, you should take mail which describes new problems and forward it to the appropriate person at Symbolics. Your Symbolics software support person will be able to tell you how to do that.

2.11.2 Backup

I can't emphasize this enough. Make sure you back up your file system regularly. Let me make that clearer. **Make sure you back up your file system regularly.**

Your file system is the repository of your work on the Lisp Machine. If your file system is damaged or destroyed, your work will be wasted.

File systems do not normally become damaged or destroyed. However, your file system can be damaged through hardware problems with your disk, software problems in the file system itself (which are hopefully rare) or in your programs, or mistakes made by users at your site (including yourself).

Backup is very cheap insurance. It takes only an hour or so a week to do routine backup of Lisp Machine file systems; at regular intervals an investment of several hours may also be necessary.

I recommend a site perform an *incremental* dump at least daily. Incremental dumps write only those files which have been modified since the last dump to tape. Depending on how much is changed on your file system, I recommend a *consolidated* dump once a week or so; these dumps contain all files modified since a specified date. I do consolidated dumps back to the last consolidated or complete dump, after which all the incremental tapes in between may be recycled. Finally, I recommend a *complete* dump when you first get your file server, and every three to six months thereafter. Keep complete backup tapes for a year or two, consolidated tapes for a year, and incrementals for a couple of weeks after they have been consolidated.

If you store your files on some other machine than a Lisp Machine file server, I still recommend you do backup. The procedures for different file systems are different; consult various systems' documentation for further information.

By the way, just doing backup is not necessarily sufficient. If you're reasonably paranoid, you'll keep at least some of your tapes someplace other than your machine room. If you have a fire or flood, not only will your machines be incapacitated, but all your work will be lost. Machines can be replaced more easily than a couple of years of your (and your colleagues') time.

2.12 Problem Set #1

This "problem set" is really just a sample programming session, to familiarize you with basic operations on the Lisp Machine.

Questions

1. Create a namespace entry for yourself.

2. Log in.

3. Switch to the editor and edit a file in your home directory named "fact.lisp."

4. Enter the text for a function named **fact** which returns

the factorial[14] of its argument.   Watch what happens
every time you type a close parenthesis. Try moving your
cursor to just after a close parenthesis on an earlier line.

5. Figure out what happens when you press Line while
   entering your function.

6. Compile the function from the editory with c-sh-C, and
   test it from the typeout window.

7. When you're satisfied with the function's performance,
   save the buffer and compile the resulting file.

8. Cold boot the machine.

9. Log in, and note that the function **fact** is now undefined.

10. Load the compiled version of your file.

11. Run your function successfully.

---

[14]Factorial of *n* is the produce of all the positive integers less than or equal to *n*.
So, (fact 4) should return 1x2x3x4=24.

Solutions

1. One way to do it: login as yourself, as in the text.

   Another way: (si:login-to-sys-host), then Edit Namespace Object. Click on [Create], click on [User], enter your chosen login-id. Fill in the required fields (marked with *), fill in any optional fields you wish. Click on [Save], and then click on [Quit].

2. Login your-user-id

3. Select E, c-X c-F fact.lisp

4. Here is one possible solution:

```
(defun fact (n)
  (if (zerop n)
      1
      (* n (fact (- n 1)))))
```

   Note the matching parenthesis blinking when your cursor is right after a close parenthesis. The same happens (in Genera 7.0) when you position the cursor over an open parenthesis.

5. Typing Line takes you down to the next line, and indents the right amount for where you are in the function. Try it in the middle of a line to see what it does.

6. Type c-sh-C while anywhere inside the text of the function to compile it. Then press Suspend to get to the typeout window, and evaluate (fact 5). The Resume key returns to the editor window.

7. Type c-X c-S, m-X Compile File (Actually, if you skip the

c-X c-S, the Compile File command will ask if you want
to save the buffer first.)

8. Press Suspend or Select L, then type Halt Machine.
   Answer the question about halting the machine Yes. Type
   B and then carriage return to the Fep Command: prompt.

9. Type Login your-user-id, then (fact 5)

10. Type Load File fact

11. Type (fact 5)

3. Flow of Control

In this chapter we temporarily leave behind the "operating system" of the lisp machine and examine aspects of the lisp language itself. In particular, we'll look at the various constructs for determining the flow of control. The notes are a little sketchier than usual, because this material is covered reasonably well in the Symbolics documentation.

3.1 Conditionals

All non-trivial lisp programs execute only those "statements," or *forms*, which are appropriate to their circumstances. They specify which forms to execute using *conditional* special operators. To test whether a condition is true, a *predicate* function is used.

Virtually all conditionals test the truth or falsehood of the conditions by comparing the result of the predicate with **nil**: if the predicate returns **nil**, the condition it is testing is false; anything else is considered to be true.[1]

Let's start with a few simple conditional operators. The **when** special operator tests the result of evaluating first "argument," which is usually a predicate form; if it is **nil**, the **when** form returns **nil** without evaluating any of the rest of its subforms. If it's not **nil**, the rest of the forms are evaluated, and the value of the last one is returned.[2] For example:

```
(when (< n 0)
  (format t "N is negative")
  n)
```

will print "N is negative" when it is, and return the value of **n**. If **n** is zero or positive, this form will return **nil**. The special operator **unless** is just like **when**, except it inverts the

---

[1]When there is no other, more obvious non-**nil** value to return, most predicates return the special symbol **t**. However, *any* value other than **nil** means "true."

[2]I have quoted the word "argument" to show the distinction between normal functions and special operators. Functions are called with arguments, which have already been passed through **eval**. Special operators, on the other hand, aren't called with all their subforms already evaluated. Rather, special operators get to decide which, if any, of their subforms get evaluated. **when**, for example, always evaluates its first subform; only if it returns some non-**nil** value do the rest of the subforms get evaluated.

result of the predicate before testing it.[3] **if** is somewhat more powerful. It evaluates its first "argument;" if its result is **nil**, it evaluates its second subform, and returns its value. Otherwise, it skips the second subform, and evaluates the *rest* of the subforms, and returns the last one's value. For example, here is one way to take the absolute value of a number:

```
(if (< n 0)
    (- n)
    n)
```

The great-granddaddy of all conditionals is called **cond**. All the others can be implemented in terms of **cond**. Each of the "arguments" (*clauses*) to **cond** is a list. The first element of the clause is a predicate (sometimes called the *antecedent*), and the rest of the list (if any) is called the *consequent(s)*. The clauses are examined one at a time, in order of appearance. If a clause's antecedent returns **nil**, the consequents are skipped and the next clause is considered. When a clause is found whose ancedent returns a non-**nil** value, that clause's consequents, if any, are all evaluated. The value of the last consequent in that clause (or of the antecedent, if there are no consequents) is the value returned by the **cond**, and the rest of

---

[3]You will sometimes see the special operators **and** and **or** misused for **when** and **unless**. For example,

```
(and bright-day
     glorious-day
     (print "It is a bright and glorious day"))
```

should really be written:

```
(when (and bright-day
           glorious-day)
    (print "It is a bright and glorious day"))
```

the clauses are skipped. If every clause's antecedent returns nil, the **cond** returns **nil.** For example:

```
(cond ((< discriminant 0)
       (error "Can't solve this quadratic."))
      ((= discriminant 0) (/ (- b) (* 2 a)))
      (t (list (/ (+ (- b) (sqrt discriminant)) (* 2 a))
               (/ (- (- b) (sqrt discriminant)) (* 2 a)))))
```

Note the use of **t** as the last antecedent of **cond.** This is a fairly common programming cliche, and means "if all else fails, this is the answer."

Other special operators based on **cond** include **case, selector** and **ecase.** Here is an example of the use of **case:**

```
(case (fruit-type fruit)
  (apple (make-apple-pie fruit))
  (orange (make-orange-marmalade fruit))
  (lemon (make-lemonade fruit))
  (otherwise (error "I don't know how to deal with ~A"
                    fruit)))
```

Look in the Symbolics documentation for more information: See the section "Conditional Functions" in *Symbolics Common Lisp: Language Concepts.*

3.2 Blocks and Exits

block is the primitive special operator for defining a piece of code which may be exited from the middle. The first "argument" must be a symbol. It is not evaluated, and becomes the *name* of the block. The rest of the "arguments" are forms to be evaluated. If a call to **return-from** occurs within the block, with a first "argument" of the block's name, the block is immediately exited. The next "argument" to **return-from** is evaluated and becomes the return value(s) for the **block**.[4]

The scope of the name of the block is lexical, so the corresponding use of **return-from** must occur textually within the block. It will not work to call **return-from** inside a function which is called within the block. The next section discusses that sort of nonlocal exit.

Blocks may be nested. That's the whole point of naming them. A **return-from** causes an immediate exit from the innermost **block** with a matching name.

Some other constructs (including **do** and **prog**) create implicit blocks. These blocks have **nil** for a name, and therefore they may be prematurely exited with (return-from nil ...). They may also be exited with the **return** special operator, which always exits the innermost block named **nil**.

Finally, certain constructs create implicit *named* blocks. Such special operators as **defun**, for example, create blocks with the obvious names.

3.3 Nonlocal Exits

catch and **throw** are analogous to **block** and **return-from**, but they are scoped dynamically rather than lexically. A **throw** may cause an exit from any **catch** on the control stack at the

---

[4]To return more than one value, say:

```
(return-from name (values v1 v2 ...))
```

time the **throw** is reached (unless an inner **catch** is shadowing an outer **catch** with the same tag). **catch**'s equivalent to the name of a block is its *tag*. The tag is the first argument to **catch**; it is evaluated, and may return any lisp object. The first argument to **throw** is its tag. It is also evaluated, and the **throw** causes the exit of the innermost **catch** whose evaluated tag is **eql** to the **throw**'s evaluated tag.

If a **throw** occurs, its second argument is evaluated and its value(s) will be returned by the corresponding **catch**. If no **throw** occurs, **catch** returns the values returned by its last subform.[5]

3.4 Iteration

There are three styles of built-in facilities for iteration. A group of operators are available for mapping a function through one or more lists; the **do** special operator allows more general forms of iteration; and the **loop** macro provides even more flexibility. This set, of course, is easily extended by writing more macros.

When more than one of the iteration facilities is applicable to a particular task, the choice is mainly a matter of personal taste.[6] All three are comparably efficient. The key issue is readability, and on this score opinions differ. My own view is that the

---

[5]There are obsolete forms of **catch** and **throw**, called *catch and *throw (both in Zetalisp only). They differ from the newer versions mainly in what values are returned. Some old system software might still use them. *catch and *throw should not be used in new code.

[6]See hacker's definition at end of chapter.

mapping functions are succinct and to the point, and therefore desirable, within the limited set of applications that are easily expressed as mapping operations. For all other kinds of iteration, **do** is often concise but I find it somewhat obscure.[7] I think **loop** is much easier to read, but there are those who consider it wordy and nebulous.

> "**loop** forms are intended to look like stylized English rather than Lisp code. There is a notably low density of parentheses, and many of the keywords are accepted in several synonymous forms to allow writing of more euphonious and grammatical English. Some find this notation verbose and distasteful, while others find it flexible and convenient. The former are invited to stick to **do**."

> [The preceding, as well as parts of the discussion of **loop** below, is taken from the loop documentation written by Glenn S. Burke: MIT Laboratory for Computer Science TM-169.]

For the sake of comparison, here are three ways to print the elements of a list:

```
(defun print-elts1 (list)          ; mapping
   (map nil #'print list))

(defun print-elts2 (list)          ; do
   (do ((l list (cdr l)))
       ((null l))
       (print (car l)))))
```

----

[7]ditto.

```
(defun print-elts3 (list)          ; loop
  (loop for elt in list
        do (print elt)))
```

3.4.1 Mapping

There are seven basic mapping functions in Common Lisp, all of which are defined on the Lisp Machine. One of them works on all kinds of Common Lisp "sequences", while the other six, inherited from earlier Lisp designs and implementations, work only on lists.

The modern, Common-Lispy way to do mapping is with the CL function **map**. It works on any sort of *sequence*, by which is meant either a list or a vector of objects. It takes the following arguments:

- *result-type* – the type of the returned value. This is usually something like **'list** or **'vector**, meaning to return a sequence of the given type; it can also be **nil**, meaning that you don't care what is returned (it returns **nil**, in fact), because you're using **map** for side-effects.

- *function* – the function to apply to the elements of the sequence(s). It should take as many arguments as the number of sequences you pass to **map**. The function can be a *lambda expression* instead of a function name. See the section "Evaluating a Function Form" in *Symbolics Common Lisp: Language Concepts*.

- *sequence(s)* – one or more arguments which are the sequences to be mapped over.

A few examples:

```
(map 'list #'+ '(1 2 3 4) #(2 4 6 8))
→ (3 6 9 12)

(map 'vector #'list #(1 2 3 4) '(2 4 6 8))
→ #((1 2) (2 4) (3 6) (4 8))

(map nil #'(lambda (a b c)
              (when (minusp (- (* b b) (* 4 a c)))
                (return-from quad-solver (values nil nil))))
     a-vector b-vector c-vector)
```

This last example either returns **nil** or returns from an outer block named **quad-solver**.

The other six mapping functions are left over from Zetalisp. There are six of them because there are two different ways the "mapped" function might be called (on elements or sublists of the input lists), and three different values they might return (a list of the values, a list formed by **nconc**ing the values, or an uninteresting value [*i.e.,* you're mapping for effect]). These functions are called **mapc**, **mapl**, **mapcar**, **mapcan**, **maplist** and **mapcon**: look them up if you want to use them.

3.4.2 Do

A **do** looks like this:

> (do ((*var init repeat*)
>   (*var init repeat*) ...)
>   (*end-test exit-form exit-form* ...)
>   *body-form body-form* ...)

The first subform is a list of iteration variable specifiers. Upon entering the **do**, each *var* is bound to the corresponding *init*. And before each subsequent iteration, *var* is set to *repeat*. The variables are all changed in parallel.

The second subform contains the *end-test* and the *exit-forms*. The *end-test* is evaluated at the beginning of each iteration. If it returns a non-nil value, the *exit-forms* are all evaluated, and the value of the last one is returned as the value of the **do**. Otherwise the *body-forms* are all evaluated. Here's an example which fills an array with zeroes:

```
(do ((i 0 (1+ i))
     (n (array-length foo-array)))
    ((= i n))
  (setf (aref foo-array i) 0))
```

Upon entry, i is bound to 0 and n is bound to the size of the array. On each iteration, i is incremented. (n stays constant because it has no *repeat* form.) When i reaches n, the **do** is exited. On each iteration, the *i*th element of foo-array is set to 0.

And another, which is equivalent to (maplist #'f x y):

```
(do ((x x (cdr x))
     (y y (cdr y))
     (z nil (cons (f x y) z)))
    ((or (null x) (null y))
     (nreverse z)))
```

Note that the preceding example has no body. It's actually fairly common for all the action in a **do** to be in the variable stepping.

There are macros named **dotimes** and **dolist** which expand into common **do** constructs. For instance, the following code macroexpands into the equivalent of the first example above:

```
(dotimes (i (array-length foo-array))
  (setf (aref foo-array i) 0))
```

3.4.3 Loop

A typical call to **loop** looks like this:

```
(loop clause
      clause
      clause ...)
```

Each clause begins with a keyword, and the contents of the rest of the clause depend on which keyword it is. Some clauses specify variable bindings and how the variables should be stepped on each iteration. Some specify actions to be taken on each iteration. Some specify exit conditions. Some control the accumulation of return values. Some are conditionals which affect other clauses. And so on. A full discussion of all the clauses would be lengthy and not particularly useful, as they're all described coherently enough in the documentation. We'll just look at some representative examples.

The **repeat** clause specifies how many times the iteration should occur. The keyword is followed by a single lisp expression, which should evaluate to an integer. And the **do** keyword is followed by any number of lisp expressions, all of which are evaluated on each iteration. So putting the two together,

```
(loop repeat 5
      do (print "hi there"))
```

prints "hi there" five times.

The most commonly used (and complicated) of the iteration-driving clauses is the **for** clause. The keyword is followed by the name of a variable which is to be stepped on each iteration, then some other stuff which somehow specifies the initial and subsequent values of the variable. Here are some examples:

```
(loop for elt in expr
      do (print elt))
```

expr is evaluated (it better return a list), **elt** is bound in turn to each element of the list, and then the loop is exited.

```
(loop for elt on expr
      do (print elt))
```

is similar, but **elt** is bound to each sublist in the list.

```
... for x = expr ...
```

expr is re-evaluated on each iteration and **x** is bound to the result (no exit specified here).

```
... for x = expr1 then expr2 ...
```

x is bound to *expr1* on the first iteration and *expr2* on all succeeding iterations.

```
... for x from expr ...
```

x is bound to *expr* (it had better return a number) on the first iteration and incremented on each succeeding iteration.

```
... for x from expr1 to expr2 ...
```

like above, but the loop is exited after **x** reaches *expr2*.

```
... for x from expr1 below expr2 ...
```

like above, but the loop is exited just before **x** reaches *expr2*.

```
... for x from expr1 to expr2 by expr3 ...
```

x is incremented by *expr3* on each iteration.

```
... for x being path ...
```

uses various other iterative constructs for looping over array elements, hast table elements, and so forth.

When there are multiple **for** clauses, the variable assignments occur sequentially by default, so one **for** clause may make use of variables bound in previous ones:

```
(loop for i below 10
      for j = (* i i) ...)
```

i starts at 0 when **from** isn't specified. Parallel assignment may be specified by using **and** instead of **for**.

The **with** clause allows you to establish temporary local variable bindings, much like the **let** special operator. It's used like this:

```
(loop with foo = expr
```

expr is evaluated only once, upon entering the loop.

A number of clauses have the effect of accumulating some sort of return value. The form

```
(loop for item in some-list
      collect (foo item))
```

or

```
(loop for item being the array-elements of some-array
      collect (foo item))
```

applies **foo** to each element of **some-list** or **some-array**, and returns a list of all the results, just like

```
(map 'list #'foo some-list)
```

and

```
(map 'list #'foo some-array)
```

The keywords **nconc** and **append** are similar, but the results are **nconc**ed or **append**ed together. Keywords for accumulating numerical results are **count**, **sum**, **maximize**, and **minimize**. All of these clauses may optionally specify a variable into which the values should be accumulated, so that it may be referenced. For instance,

```
(loop for x in list-of-frobs
      count t into count-var   ;"t" means always count
      sum x into sum-var
      finally (return (/ sum-var count-var)))
```

computes the average of the entries in the list.

The **while** and **until** clauses specify explicit end-tests for terminating the loop (beyond those which may be implicit in **for** clauses). Either is followed by an arbitrary expression which is evaluated on each iteration. The loop is exited immediately if the expression returns the appropriate value (nil for **while**, non-nil for **until**).

```
(loop for char = (read-char *standard-input*)
      until (char-equal char #\end)
      do (process-char char))
```

The **when** and **unless** clauses conditionalize the execution of the following clause, which will often be a **do** clause or one of the value accumulating clauses. Multiple clauses may be conditionalized together with the **and** keyword, and *if-then-else* constructs may be created with the **else** keyword.

```
(loop for i from a to b
      when (oddp i)
        collect i into odd-numbers and do (print i)
      else collect i into even-numbers
      finally (return (values odd-numbers
                              even-numbers)))
```

The **return** clause causes immediate termination of the loop, with a return value as specified:

```
(loop for char = (read-char *standard-input*)
      when (char-equal char #\end)
        return "end of input"
      do (process-char char))
```

Please refer to the documentation for a more complete discussion of **loop** features. Of particular importance are *prologue* and *epilogue* code (the **initially** and **finally** keywords), the distinction between ways of terminating the loop which execute the epilogue code and those which skip it, the way to name the body of the loop for returns (**named**), aggregated boolean tests (the **always, never,** and **thereis** keywords), the destructuring facility, and iteration paths (user-definable iteration-driving clauses).

3.4.4 Implicit Iteration

No discussion of iteration would be complete without mentioning that many of the usual iterative functions you want to use are already present in the language. There are certain iterative constructs that *everybody* wants sometime in their programming lives, and are so commonly used that they have become part of the Lisp language. Some examples:

- **remove** – drop elements of a sequence when they fit some criterion.

- **union** – create a sequence which contains all the elements of two or more sequences.

- **intersection** – create a sequence which contains all the elements which two or more sequences have in common.

- **reverse** – create a sequence which contains the original elements in the opposite order.

- **find** – search for an element of a sequence which meets some criterion.

These have all been adopted by Common Lisp, and are implemented on the Lisp Machine. See the section "Sequence

Operations" in *Symbolics Common Lisp: Language Concepts*.
Other examples of implicit iterative constructs include
do-symbols and **maphash**.

One other "iterative" function deserves mention here. The
sort function takes any sequence and puts it in order, according
to some sorting predicate you supply. For example, if you have
a list of strings you want sorted alphabetically, here is one way
to do it:

```
(setq strings (sort strings #'string-lessp))
```

3.5 Lexical Scoping

Whenever two different functions use the same name for a
variable, a conflict potentially occurs between them. Lexical
scoping helps keep the conflicts down to a manageable level.

Consider this example, taken from the Symbolics documentation:

```
(defun my-mapc (funct list)
  (loop for x in list do          ; x is bound here
    (funcall funct x)))

(defun print-long-strings (strings x) ; x is bound here
  (my-mapc #'(lambda (str)
              (if (> (length str) x) ; which x is this?
                  (print str)))
          strings))
```

If the **x** from **my-mapc** is used in evaluating the function, the
wrong result is assured. In order to keep programmers from
having to know everything about each other's functions, *Lexical
Scoping* was adopted. Under lexical scoping, names refer to the

variable whose binding surrounds the point at which they are referenced. So, for example, in the definition of **print-long-strings** above, the **x** in the internal function refers to the **x** bound by the function **print-long-strings**. Lexical scoping is usually what you want.

In addition to variables, other names are also lexically scoped. These include function names (internal functions created using **flet**, **labels** and **macrolet**) and block names (created with **block**). Catch tags (**catch**) are explictly *not* lexically scoped.

It is possible to force the Lisp Machine to use the "other" type of variable scoping, called *Dynamic Scoping*. To do this, declare the variable to be *special*. This can be done by using the **declare** special operator, or by creating the variable with **defvar** or **defparameter**.[8]

3.6 Macros

Macros are programmer-supplied extensions to the Lisp language. They are special functions which the compiler and the evaluator call to translate the source program they read into the program they eventually compile or evaluate. Macros are probably best explained by example.

Suppose you were writing a program which tested numbers for, say, being odd integers, and discovered that you were writing the following all over your program:

---

[8]There is yet a third kind of scoping, which might be called "flavor scoping," which is used for instance variables and functions declared with **defun-in-flavor**. Inside the lexical contour of a method or flavor function, for example, variable names are checked to see if they might be instance variables before looking for global definitions. This is a special case of lexical scoping. See chapter 5.

```
(when (and (integerp <variable>) (oddp <variable>))
   (setq <variable> (- <variable> 1))
   <do-something-or-other>)
```

After a while, you might like to do the following instead:

```
(with-odd-integer-rounded-down <variable>
   <do-something-or-other>)
```

This is one way you might do this:[9]

```
(defmacro with-odd-integer-rounded-down (var &body body)
   '(when (and (integerp ,var) (oddp ,var))
      (setq ,var (- ,var 1))
      ,@body))
```

Now, I admit this is a pretty simple-minded example. You're unlikely to write such simple macros. However, in cases where form you're trying to create is

1. long,
2. tedious to type,
3. error-prone, or
4. used in a lot of places in your program,

a macro can be a lifesaver.

Some hints for debugging your macros:

An easy way to expand macro forms in a lisp listener is with the CP command Show Expanded Lisp Code, or with the function **mexp**. The former takes the macro form as an argument, and a lot of keywords which allow you to tailor the exact level of expansion. For example, certain things which are defined as special operators for the interpreter are defined as macros for

---

[9] I will not be documenting how to write macros here. That is adequately covered in the Symbolics documentation. See the section "Macros" in *Symbolics Common Lisp: Language Concepts*.

the compiler; this command allows you to specify whether you want the interpreter expansion or the compiler expansion.

mexp enters a loop which reads forms from the keyboard, expands them, and prints the result, using the pretty-printer for readability. If you want to see how a macro expands when given different sets of arguments, this is often a good way to find out what they all do.

In the editor, a useful command to know is control-shift-M. With the cursor positioned immediately *before* the form you wish to expand, type c-sh-M. The editor will display the macro expansion in the typeout window. If you would like to have the macroexpansion inserted in the buffer, use c-sh-M with a numeric argument.

c-sh-M only expands a given form once. Sometimes macros expand your code into code which in turn uses another macro! To see the ultimate code the compiler or interpreter will see, type m-sh-M. This macroexpands your input form until it contains no more macros to expand. m-sh-M expands subforms as well.

Of course, you can also use meta-. on a macro to edit its definition, but it's often more useful to see what a macro expands into than to see how it's implemented.

3.7 Unwind-protect

One particular special operator needs to be mentioned, because it is so useful, especially in conjunction with macros. The operator **unwind-protect** is used to make sure you get a chance to clean up after yourself, even if the function you're running doesn't return normally.

Consider the following code fragment:

```
(defun operate-reactor (reactor)
  (pull-out-control-rods reactor)
  (generate-electricity reactor)
  (push-in-control-rods reactor))
```

Now, suppose there is an error in your function **generate-electricity**, and it enters the debugger. You poke around, figure out the problem, and want to try it again. If you type control-Abort, the last form in this function *never gets evaluated.* You could wind up blowing up an area the size of the state of Pennsylvania.

The way to make sure you get to clean up is with the special operator **unwind-protect**. Here is the above form, recoded to use it:

```
(defun operate-reactor (reactor)
  (unwind-protect
      (progn (pull-out-control-rods reactor)
             (generate-electricity reactor))
    (push-in-control-rods)))
```

The first subform in the **unwind-protect** is called the *protected form*. It is executed in an environment which guarantees that all the remaining forms, called the *cleanup forms*, will be executed when the protected form returns, or is aborted using **return**, **return-from** or **throw**.[10] The **progn** special operator is how you make several forms into a single one; **unwind-protect** is one of the few cases where a single form is required (if is another).

---

[10]Control-Abort is implemented using **throw**.

Now, if you need to do this in more than one place in your program, you will appreciate the following macro:

```
(defmacro while-reactor-running ((reactor) &body body)
  '(unwind-protect (progn (pull-out-control-rods ,reactor)
                          ,@body)
     (push-in-control-rods ,reactor)))

(defun operate-reactor (reactor)
  (while-reactor-running (reactor)
    (generate-electricty reactor)))
```

A couple of examples of system macros which use **unwind-protect**: **with-open-file** and **with-open-stream**.

3.8 Fun and Games

From *The Hacker's Dictionary*, Guy L. Steele, Jr., *et al*:

TASTE *noun.*   Aesthetic pleasance; the quality in programs which tends to be inversely proportional to the number of FEATURES, HACKS, CROCKS, and KLUGES programmed into it.

OBSCURE *adjective.*   Little-known; incomprehensible; undocumented. This word is used, in an exaggeration of its normal meaning, to imply a total lack of comprehensibility. "The reason for that last CRASH is obscure." "That program has a very obscure command syntax." "This KLUDGE works by taking advantage of an obscure FEATURE in TECO." The phrase "moderately obscure" implies that it could be figured out but probably isn't worth the trouble.

3.9 Problem Set

Questions

1. Write a function which takes a string as an argument, and returns its position in the list ***my-strings***. Try doing this with **do**, **loop** and **map** (**map** requires the most ingenuity). Your function should return 0 if the string is the first element in the list, 1 if it's the second, and so forth. If the string is not in the list, your function should return **nil**.

2. Write a predicate which takes two strings and returns **t** if the first one is earlier in the list ***my-strings***. If either string is not in ***my-strings***, it is later; if both are not in the list, return **t** if the first string is alphabetically earlier than the second.

3. Write a function which reads characters from the user one at a time, using **read-char**. When the user types the #\return character, it should take the characters accumulated thus far, make them into a string, and add them to the end of the list ***my-strings***. When the user types the #\end character, it should return. Do this two ways: one using **string-append**, and the other more efficiently using storage by pre-allocating the string.

4. Write a function like the last one, except that if the user presses Abort no changes are made to the list ***my-strings***. Do this in two ways: use **let** and **unwind-protect**.

5. Write a function which takes two arguments, a number *n* and a list of numbers *list*, and returns a list consisting of the elements of *list* incremented by *n*.

6. Write a function which takes one argument, n, and returns the nth Fibonacci number.[11] The only restriction is that the time the function takes should increase linearly in n (it's easy to make a recursive one which behaves much worse!).

---

[11]The Fibonacci sequence of numbers starts with 1, 1, 2, 3, 5, 8 ...; each Fibonacci number after the first two numbers is the sum of the two preceding ones.

Answers

1. Three different solutions, using **do**, **loop** and **map**. Note that this function is equivalent to using the **position** function.

```
(defvar *my-strings* '("This" "is" "the" "cereal"
                       "that's" "shot" "from" "guns"))

(defun find-string-do (string)
  (do ((string-list *my-strings* (cdr string-list))
       (n 0 (1+ n)))
      ((null string-list) nil)
    (when (string-equal (car string-list) string)
      (return n))))

(defun find-string-loop (string)
  (loop for my-string in *my-strings*
        as n upfrom 0
        when (string-equal string my-string)
          return n))

(defun find-string-map (string)
  (let ((n 0))
    (map nil (lambda (my-string)
               (when (string-equal string my-string)
                 (return-from find-string n))
               (incf n))
         *my-strings*)))
```

2. This one assumes one of the previous three functions was named **find-string**.

```
(defun my-string-lessp (string1 string2)
  (let ((index1 (find-string string1))
```

```
          (index2 (find-string string2)))
   (cond ((and index1 index2) (< index1 index2))
          (index1 t)
          (index2 nil)
          (t (string-lessp string1 string2)))))
```

3. The second of these replaces **string-append** with **vector-push-extend,** and **append** with **nconc.** I suggest you look these up if you don't know the difference between them.

```
(defun read-my-strings-slow ()
  (loop with string = ""
        for char = (read-char)
        when (char= char #\end) return nil
        when (char= char #\return)
          do (setq *my-strings*
                   (append *my-strings* (list string))
                   string "")
        else do (setq string
                      (string-append string char))))

(defun read-my-strings-fast ()
  (loop with string = (make-array
                        10
                        :element-type 'string-char
                        :fill-pointer 0)
        for char = (read-char)
        when (char= char #\end) return nil
        when (char= char #\return)
          do (setq *my-strings*
                   (nconc *my-strings* (list string))
                   string (make-array
                           10
                           :element-type 'string-char
```

```
                             :fill-pointer 0))
             else do (vector-push-extend char string)))
```

One problem with these functions is that if you make a mistake, you're dead; no provision has been made for rubbing out errors. Try modifying the second one to handle the character #\rubout.

4. Neither of these functions is very clear as to what it does. A better solution will be shown below.

```
(defun read-my-strings-protected ()
  (let ((result *my-strings*))
    (let ((*my-strings* (copy-list *my-strings*)))
      (read-my-strings)
      (setq result *my-strings*))
    (setq *my-strings* result)))

(defun read-my-strings-protected ()
  (let ((old-strings (copy-list *my-strings*)))
    (unwind-protect
        (progn (read-my-strings)
               (setq old-strings *my-strings*))
      (setq *my-strings* old-strings))))
```

The **copy-list** calls are only necessary if you use the version of **read-my-strings** which uses **nconc** instead of append.

A more perspicuous way:

```
(defun read-some-strings ()
  (loop with string = ""
        for char = (read-char)
        when (char= char #\end) return nil
```

```
            when (char= char #\return)
              collect (prog1 string (setq string ""))
            else do (setq string
                             (string-append string char)))))

(defun read-my-strings-protected ()
  (let ((new-strings (read-some-strings)))
    (setq *my-strings*
          (nconc *my-strings* new-strings))))
```

This version doesn't have any complicated interactions between binding variables and exiting from lexical contexts unexpectedly, and therefore is easier to understand.

5. This one works only on lists:

```
(defun add-to-sequence (n list)
  (map 'list (lambda (elem) (+ n elem)) list))
```

This one works on all Common Lisp sequences (vectors as well as lists):

```
(defun add-to-sequence (n seq)
  (map (if (listp seq) 'list (type-of seq))
       (lambda (elem) (+ n elem))
       seq))
```

6. Here are a couple of different solutions.

```
(defun fib (n)
  (loop repeat n
        for a = 0 then b
        for b = 1 then partial-sum
        for partial-sum = (+ a b)
        finally (return partial-sum)))
```

Here is a completely different way to do it:

```
(defun make-fibber (a b)
  #'(lambda ()
      (psetf a b b (+ a b))
      a))

(defun fib (n)
  (let ((fibber (make-fibber 0 1)))
    (loop repeat n do (funcall fibber)
          finally (return (funcall fibber)))))
```

The function **make-fibber** returns a *lexical closure*, which is a function that remembers the lexical environment in which it was created. Each time you call the function returned (the "fibber"), it uses the values of **a** and **b** as they were the last time that lexical environment was entered. This kind of function is sometimes called a generator function.

It is possible to make closures of more than one function in the same lexical environment, in which case whenever any of them is invoked, it sees the variables in the state they were in the last time any of them was invoked. For example:

```
(defun make-cache-functions (cache-value)
  (let ((cache-count 0))
    (values #'(lambda () (values cache-value
                                 cache-count))
            #'(lambda (new-value)
                (incf cache-count)
                (setf cache-value new-value)))))

(defun cache-user ()
```

```
(multiple-value-bind (get set)
    (make-cache-functions 1)
  (list (funcall get)
        (progn (funcall set 105) (funcall get)))))
```

This simple function returns a list with two elements: 1 and 105.

4. More on Navigating the Lisp Machine

The last chapter discussed aspects of programming with the Lisp language. This one is about some aspects of using the Lisp Machine which are more or less independent of programming on it, *i.e.*, what you might call the operating system of the Lisp Machine. Some parts of this chapter having to do with sending messages to windows and process objects may be confusing unless you know something about flavors. See chapter 5, especially the section starting at page 117.

4.1 The Scheduler and Processes

A *process* is a single computational sequence within a computer. The Lisp Machine supports multiple processes running "simultaneously," *i.e.*, sharing the processor like a miniature time-sharing system. Each process behaves like it has its own simulated processor: it has its own "program counter," its own function-call history (*stack*), and its own special-variable bindings.

Switching the processor back and forth among the different

processes can be explicitly controlled by the Lisp Machine programmer (read the documentation on *Stack Groups*), but almost never is. A special module called the *scheduler* generally handles this responsibility. Every 1/60th second the scheduler wakes up and decides whether the current process should be allowed to continue running, and if not, which other process should get a chance.

If the current process has been running continuously for less than a second, and wishes to continue, it is allowed to. (Note that a full second is a long time for this sort of thing, compared to other timesharing arrangements.) Or if it's been running for a second but no other process wishes to run, it is still allowed to continue. But if it's been monopolizing the machine for more than a second, and one or more other processes want to run, it's forced to take a rest while the scheduler gives the others a chance. The process chosen by the scheduler is now treated as the previous current process was: it will be allowed to run until some other process(es) wish to run and the current process either volunteers to give the others a chance, or passes the one second mark.[1]

The way a process "volunteers to give the others a chance," or, in less emotionally-laden terms, informs the scheduler that it doesn't need to run, is with the function **process-wait**. The function which calls **process-wait** specifies a condition the process is waiting for. When the condition becomes true, the process is ready to run. When the scheduler decides to resume the process, the call to **process-wait** returns and the computation continues from there. The first argument to **process-wait**

---

[1]The "one second" referred to is actually a shorthand way of saying "the process' quantum." The default runtime quantum is one second, but may be modified for any process. See the section "Process Attribute Messages" in *Internals, Processes, and Storage Management*.

is a string to appear in the "wholine" (at the bottom of the screen) while the process is waiting. The second argument is a function and any remaining arguments are arguments to the function. To see whether the process is ready to continue, the scheduler applies the specified function to the specified arguments. The return value of the function is what the scheduler uses for the "condition" mentioned above. This function is often called the process' *Wait Function.*

Here is a simplified version of the call to **process-wait** which is responsible for "User Input" appearing in the wholine most of the time:

```
(process-wait si:*whostate-awaiting-user-input*
  #'(lambda (buffer) (not (io-buffer-empty-p buffer)))
  buffer)
```

This call is buried somewhere in the code windows use for reading from the keyboard. It says that the process will be ready to continue when the function **io-buffer-empty-p** returns **nil** when applied to the input buffer.

Suppose several processes' wait functions would all return non-**nil** values at a given moment. Which process gets to run next? The scheduler orders all processes in a list by their *priorities*; the highest-priority processes' wait functions are checked first. The priority of a process is set when you create the process, although it can be changed by sending the process object a **:set-priority** message.

Now a question for the bold: what happens if an error occurs in the scheduler? It is, after all, just another piece of Lisp code. And even if the scheduler code itself is bug-free, all the wait-functions are called in the scheduler, and any loser[2] can

---

[2]See hacker's definition at the end of the chapter.

write a buggy wait-function. Blinking of flashing blinkers also gets done from the scheduler. (There's a *clock function list* of things to be done every time the scheduler runs, and by default the only things on the list are blinking the blinkers and updating the mouse documentation on the screen.) And any loser can also write a buggy :blink method for his/her blinkers – I've certainly done it. So what happens when the scheduler runs into an error? The scheduler can't use the Window System, since any window it might use could be locked. How can the debugger communicate with you?

What happens is that the scheduler enters the debugger and uses what is called the *cold-load stream*. This is a very basic stream which completely bypasses the Window System. It uses the screen as it would a dumb terminal, with no regard for the previous display contents, ignoring even window boundaries. None of the input editor commands will work, apart from the rubout and Clear-Input keys. Things like Control-Abort won't work. But you will be in a legitimate debugger, from which you can attempt to set things right. So don't panic.

Our view of scheduling is now fairly complete. The current process owns the Lisp Machine until it either does a **process-wait**, or uses up its second. When either of these occurs, the scheduler calls the wait-functions of the other processes. The first process whose wait-function returns a non-**nil** value gets to become the ***current-process***. If none of them do, the old current process remains the current process. And if any errors occur while in the scheduler, the debugger uses the cold-load stream.

Fine. Now it's time to complicate things again. At any given time a process is either *active* or *inactive*. Inactive processes are not even considered by the scheduler when it looks for an alternative to the current process. Their wait-functions aren't called at all until they become active. And what makes a process active or inactive? Two of the instance variables of a process are its *run-reasons* and its *arrest-reasons*. An active process is one with no arrest reasons and at least one run reason. Otherwise (at least one arrest reason or no run reasons) the process is inactive. There are program interfaces for looking at a process' run and arrest reasons, and for adding to or deleting from them. An interactive user, however, is more likely to arrest or un-arrest a process in one of the following ways (all of which end up calling those same interfaces, but are easier to use from the console):

1. The System Menu has options for arresting or un-arresting the process in the window the mouse is over.

2. If you click on the name of a process in Peek's display of processes, you get a menu of useful things to do to that process. The menu includes "arrest" and "un-arrest" options.

3. Clicking right on the name of a process printed by the Show Processes command will also provide a menu, this time of CP commands which have to do with processes.

4. Typing Function A arrests the process the wholine is watching. (This is usually the selected window's process. You can change which process the wholine watches with Function W.) Function minus A un-arrests it.

5. Typing Function Control-A arrests all the processes except for a few without which you couldn't run the machine, like the keyboard and mouse processes. You can then use Function minus A on specific processes, or Function minus Control-A to turn them all back on.

Another common operation to perform on a process is to *reset* it. This is very much like typing c-m-Abort to it. It flushes everything on the process' stack and restarts it. (More exactly, it reapplies the process' initial function to its initial arguments, but you needn't understand that just yet.) The only time you can type c-m-Abort to a process when you can select its window, which isn't always possible, but you can reset a process anytime. The options for how to reset a process are similar to those for un-/arresting one. You can use the [Reset] option in the system menu to reset the process in the window under the mouse, or you can use the menu in Peek's display of processes. A CP command, Restart Process, is also provided. Finally, there is also a programmer's interface for resetting a process.

Note that most of these ways of resetting depend on being able to use the mouse. So if the mouse process is the one which is in trouble, they won't work. The CP command Initialize Mouse (or the function **tv:mouse-initialize**) are provided for this contingency.

One final note on resetting: (send *current-process* :reset) doesn't work. (It just returns **nil**.) The usual method for unwinding a stack doesn't work from within that stack. To reset the current process, you need to either spawn a new process for the sole purpose of resetting your process (use **process-run-function**), or use an optional argument to the reset message: (send *current-process* :reset :always) will work.

Before long you will probably have cause to create your own processes. The easy way to do this is with the function **process-run-function**. See the section "Creating a Process" in *Internals, Processes, and Storage Management*.

4.2 Windows

Processes usually communicate with the user through one or more *windows*. A window is a rectangular piece of the screen, which displays output sent to it, and obtains its input via the keyboard and mouse.[3]

The entire set of existing windows is organized into several trees. The root of each tree is a *screen*, which is a software representation of a "display." Each window has a *superior* (towards the root of the tree) and a (possibly empty) list of inferiors (towards the leaves). The [Windows] option in Peek displays all the trees (subject to a restriction mentioned below).

A window may be in one of four states:

1. Deactivated
2. Deexposed
3. Exposed
4. Selected

A newly-created window starts off in the *deactivated* state. The window system doesn't remember deactivated windows at all, so if you don't keep a pointer to it, its storage can be reused by the garbage collector.[4] (Ignore this point for now if you don't understand garbage collection.)

---

[3]The programmer's interface to the window system uses Flavors. Unfortunately, the implementation of windows is much older than the current Flavors system, and uses "message passing" interfaces. For further information: See the section "Message Passing," page 117.

[4]Normally, the window system remembers a window in two places: (1) the window's superior remembers all its inferiors, and (2) an array called **tv:previously-selected-windows** remembers all windows which have been selected.

In order to use a window at all, you must *activate* it; this usually happens automatically as part of exposing it on the screen for the first time. *Exposing* a window means making it completely visible on its superior. Since most windows you will create will have a screen as their superiors, exposure means that the window is visible on the screen. Note that it is possible for a window to be partially visible, because some other window is covering up part of it; a partially visible window is *not* exposed.

In order to write output to a window, the output must have someplace to go. For exposed windows, this "someplace to go" is the window's superior, *e.g.*, the screen. A deexposed window (including one which is partially visible) can have a bit-array, called its *bit-save array*, which can be copied to the screen later. See the section "Pixels and Bit-Save Arrays" in *Programming the User Interface, Volume B*.

A deexposed window which is asked to display output (perhaps with the **write-char** or **graphics:draw-rectangle** functions), has several options. What happens is controlled by its *deexposed typeout action*. It may specify, for instance, that the window should try to expose itself, or that an error should be signaled. The default value of deexposed-typeout-action, **:normal**, specifies that the process doing the typeout should enter an *output hold* state. That means it will do a **process-wait** (remember those?) with a wholine state of "Output Hold" and a wait-function which essentially waits for the window to become exposed:

```
(process-wait "Output Hold"
  #'(lambda (sheet)
      (not (sheet-output-held-p sheet)))
  self)
```

You can also set the deexposed typeout action of a window to **:permit**, either by sending it the appropriate message or by using Function 4 T or Function 5 T. In this case, output to the window will proceed (if it has a bit-save array), but you won't be able to tell unless you expose it.[5]

So much for output. Suppose you have two exposed windows on your screen, and each has a process which is waiting for typed input. How does the Lisp Machine decide which one should get the characters you type?

The answer is *selection*. The *selected window* is the one to which keyboard input is directed. Although any number of windows may be simultaneously exposed, as long as they can all fit on your screen without overlapping, only one window at a time may be selected. The currently selected window is always the value of the symbol **tv:selected-window**. It usually has a blinking rectangular cursor in it.[6]

If we imagine the four possible window states (deactivated, deexposed, exposed, selected) occupying a spectrum, the various operations for changing the state of window are pictured in Figure 1. The vertical lines show the transitions between the various states. The arrows mean that when you perform the given operation on a window, its state gets pushed all the way from where it is to the head of the arrow. So, for instance, if you deactivate an exposed window, it will be both deexposed and deactivated. A selected window would be deselected, deexposed, and deactivated. The operations only push in the direc-

---

[5]You might like a partially exposed window's contents to be updated on the screen as output occurs. See the variable **tv:screen-manage-update-permitted-windows** in *Programming the User Interface, Volume B*.

[6]There must, of course, be a process "listening" for input to the selected window for keyboard input to have any effect. Otherwise, typing on the keyboard has no effect, except for special keys like c-Abort, Function and Select.

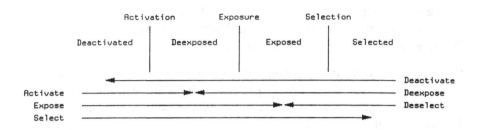

Figure 1.   Transitions among window states

tion of the arrows, they don't pull. That is, if the window is already at or beyond the arrowhead nothing happens. If a selected window is activated, there is no effect. It is not pulled back to the deexposed state.

A freshly instantiated window returned by **tv:make-window** will be deactivated, unless you specify otherwise. This is also the state of a window which has been explicitly deactivated or killed. (Killing a window deactivates all of its inferiors as well as itself.)

You can always change the state of a window by sending it an appropriate message, but there are several ways to make these messages be sent without explicitly sending them yourself. The system menu has an option for killing the window under the mouse, and one for selecting a window from the list in **tv:previously-selected-windows**. The [Edit Screen] option in the System Menu pops up another menu with options for killing or exposing any partially visible window, and for exposing any window in **tv:previously-selected-windows**. The Edit Screen menu also has options for creating, moving, or reshaping windows. If you click on the name of a window while in the windows display of Peek, you get a menu with options for selecting, deselecting, exposing, deexposing, deactivating or killing the window.

There's another way to select a window which you are already familiar with: use the Select key. For the kinds of windows accessible via the Select key (Select Help displays a list), the effect of the Select key depends on how many instances of that flavor of window exist.

Let's take Select L (for the Lisp Listener) as an example. If there are no existing LL windows, typing Select L will create one and select it. If there is exactly one LL window, Select L will select it (unless it is already the selected window, in which case the console will beep and the window will remain the selected window). If there are more than one existing LL windows, and none of them are the selected window, Select L will select the one which had most recently been the selected window. Typing Select L repeatedly will rotate through all the existing Lisp Listeners.

Typing Select c-L (hold down the control key while striking L) will always create and select a new Lisp Listener window, regardless of whether there any already exist.

Windows can also be selected with the Function key. Function S selects the previously selected window. Typing a numeric argument (pressing digit characters between the Function key and the S) allows rotation of the selected windows in various arcane ways. Type Function Help and read about Function S for a full description. [In addition to the usual ways of exposing the window, when an output hold occurs there is one extra way which becomes available: type Function Escape.]

Changing the state of a window will often cause the state of other windows to change. For instance, if I select one window, the window which had been selected necessarily becomes deselected. And if I deselect a window, some other window (the previously selected one) becomes selected. Similarly, exposing a window may partially or entirely cover some other window which had been exposed; the latter window is forced to become deexposed. And deexposing a window may uncover some other window, thereby exposing it.

(A subtler point arises here. Simply sending the **:deexpose** message usually does not have the intended effect. Since no other windows will be covering the one which has just been deexposed, it will immediately be automatically re-exposed. It will look like nothing happened. What you probably meant to do was either expose some other window [which will automatically deexpose the first window], or send the first window the **:bury** message, which in addition to deexposing it, puts the window underneath all the other windows, so that the window that ends up being auto-exposed is some other one. Function B can be used to bury the selected window.)

The interactions among windows can become terribly convoluted. There are several kinds of locks intended to keep everything straight. If something goes wrong and an error occurs while the Window System is locked, the debugger won't be able to expose a window to use. So it uses the cold-load stream, just as when an error occurs inside the scheduler.

If you've been messing with the Window System in unwise ways, it's possible to get it locked up so that you can't do anything. (I do it all the time). If the window which appears to be selected isn't responding to typein, and c-m-Abort doesn't help, and the mouse is dead, and you can't select some other window with the Select or Function keys, it may be that you're hung up in a locked Window System. Your last resort in such a case (short of h-c-Function and a warm or cold boot) is to type Function c-Clear-Input. This clears all the locks in the Window System. It's a sledgehammer, and can break some things, but it may revive your machine without having to boot.

When a window is sent more than one screenful of typeout at a time, it may pause at the end of each screenful, type **MORE**, and wait for you to press any key before continuing. This behavior is called *more processing*. Whether more processing occurs (as opposed to continuous output) is controlled by Function M and Function c-M. Type Function Help for details. For more processing to occur it must be turned on both globally and for the individual window.

One last note about navigating around the Window System. Many ways have been provided to get to other windows, not all of which are immediately obvious. Here is a list of things which you might want to try:

Select Key: Select Help -- A "Cheat sheet"

Function Key: Function Help -- A "Cheat sheet"
               Function S -- select another window
               Function B -- bury the current window,
                             selecting a new one
               Function M -- modify **More** processing
               Function T -- toggle deexposed-typeout-action
                             and deexposed-typein-action.

Certain special chararacters can be typed using the Symbol modifier key. Type Symbol-Help (*i.e.,* hold down Symbol while typing Help) for a list of them.

4.3 Debugging

The Lisp Machine provides a number of tools to help you debug
your programs. These include:

- a compiler which diagnoses many of the more obvious errors

- a powerful debugger which permits examination of your
  execution state

- dynamic breakpoints

- a monitor facility which can automatically interrupt your
  program when it sets or references a given variable

- source locators: when your program is interrupted, you
  can see what place in your source is executing

- bug report mail

The debugger itself is the main tool for discovering what went
wrong. You should remember a few things about the debugger
from Chapter 2: it is entered whenever an error occurs, and
may be entered manually with the function **break** or by typing
c-m-Suspend. Breakpoints and references to monitored variables
(see below) can also cause your process to enter the debugger.
Once in the debugger, you can move up and down the stack
with c-P and c-N (for *previous* and *next*), and see the whole
stack with c-B (for *backtrace*). There are generally a series of
restart and other commands bound to the *super* keys, and to
Resume and Abort.

Now some debugger facilities that may be new. The local variables in the current frame, including the arguments, are accessible by typing their names. Suppose the current function has

an argument named **array** and I want to know what element #5 in the array is. I could type (aref array 5), just as it might appear in my source for the function.[7] All symbols are interpreted as if they appeared in your source; variables, for example, are found using the normal lexical scoping rules. Any modifications to local variables will be used if you type Resume.

You can also examine the values of arguments and local variables with the functions **dbg:arg** and **dbg:loc**. These two functions take a single argument, which is either an integer or the name of the variable (remember that arguments are numbered from 0, not one! **(dbg:arg 0)** is the first argument.) Also, the debugger commands c-m-A and c-m-L can be used for printing out the contents of arguments and locals, respectively. c-2 c-m-A prints the third argument, setting the variable * to it.

Suppose you're debugging some function which is part of an enormous program. Your program has just spent the last 15 hours getting to the place where it blew up, and rather than restarting the computation from the beginning, you would like to continue from where you are.

If you have recompiled a function or one of its subroutines, you can *reinvoke* it, that is, start it from the beginning, using the debugger comand :Reinvoke. This can also be typed as a single character, namely c-m-R. You can also change the values of the arguments for reinvoking the function in one of two ways:

1. Modify the arguments by incanting:

---

[7]In fact, you can copy forms from the editor by pushing them onto the "kill ring" and "yanking" them into the debugger, as long as you are careful about packages (if you're not using packages, you won't need to worry about this.) See the section "The Input Editor and Histories," page 83.

(setf (dbg:arg *n*) *new-value*)

and then reinvoke the function with c-m-R. [Remember that arguments are numbered from zero, not one! *n* can be an argument number or the symbol which names the variable.]

2. Use the debugger command :Reinvoke :New Args.[8] c-m-R with a numeric argument is an abbreviation for this command.

You may find a few other debugger commands invaluable:

- c-E — takes you to the editor and positions the cursor at the place where your function halted.[9]

- c-M — takes you to a mail-sending window with the contents of the stack in the mail buffer. You can use this to send bug reports to the maintainer of whatever software you're running (including Symbolics!).

- c-m-Z — the Analyze Frame command, which tries to figure out what went wrong with your function. Usually it will tell you things like what argument was probably passed in wrong, and so forth.

Monitoring variables is a very powerful way to determine what's wrong with a program. For example, suppose you've

---

[8]This is broken in Genera 7.0; it will be fixed in a later release.

[9]For this to work completely correctly, you must compile your code with *Source Locators*. Currently, the only way to get source locators is with the editor's compilation commands. By default, the command c-m-sh-C compiles with source locators, and c-sh-C compiles without. You can reverse these by setting **compiler:*inhibit-using-source-locators*** to **nil** in your init file.

been staring at a program for a week, trying to figure out when an instance variable gets set to **nil** instead of a number. Instead of continuing to stare at it, you can merely monitor the variable.

The simplest way to monitor a variable is to use the macro **dbg:monitor-variable**. You can also explictly monitor instance variables with **dbg:monitor-instance-variable**. Examples:

```
(dbg:monitor-variable *foo*)
```

Any attempt to write the variable ***foo*** will invoke the debugger. If you press Resume, the program will continue.

```
(dbg:monitor-instance-variable *pie* 'spice-ingredients
                               :read t)
```

If your program tries to read the instance variable **spice-ingredients** of the instance ***pie***, the debugger will be entered.

```
(dbg:monitor-variable (pie-spice-ingredients *pie*) :read t)
```

will have the same effect as the previous example, if you have declared the instance variable **spice-ingredients** to be *locatable* (See the special form **defflavor** in *Symbolics Common Lisp: Language Concepts.*)

4.4 The Input Editor and Histories

The *input editor* is active in most contexts outside of the editor. Most notably, it is active when you're typing to a Lisp Listener. c-Help lists all the input editor commands. Most of them are similar to the Zmacs commands, so you can do all sorts of editing of the input before it gets to the Lisp reader. Two of the helpful features of the input editor are the *histories* it keeps, the *input history* and the *kill history*. Every time you send a form off to be evaluated by a Lisp Listener, the form is added to that Lisp Listener's input history. (Each Lisp Listener keeps its own input history, even the editor's typeout window.) Pressing the Escape key will display the input history of the window you are typing to.

Every time you delete more than one character of text with a single command (with, for example, m-D, m-Rubout, Clear-Input, c-W, c-K), the deleted text is added to the kill history. There is only one kill history; it is shared by all the windows which use the input editor, and also the Zmacs window(s). c-Escape displays the kill history.

In both the input editor and in Zmacs, c-Y *yanks* the most recent item off the kill history and inserts it at the current cursor position. You can select an earlier element from the history by giving c-Y a numeric argument. Typing m-Y immediately after a c-Y does a *yank pop*; it replaces the text which has just been yanked with the previous element from the history. Repeatedly typing m-Y will rotate all the way through the history. Giving m-Y a numeric argument will jump that many items in the history.

Note that since all windows share the same kill history, it provides a simple way to transfer text from the editor into a Lisp Listener: just push the text onto the kill history while in the editor, perhaps with c-W or m-W or a mouse command. Then switch to a Lisp Listener, type c-Y, and presto! There's your text.

In the input editor, c-m-Y yanks from the input history. m-Y again has the effect of rotating through the history. Only input which is appropriate to the *input context* is actually yanked.

So, for example, if you're supposed to be typing a command, then c-m-Y and m-Y will yank commands you previously typed to the window. If you've already typed a command name, and the command is prompting you for a file pathname, c-m-Y/m-Y will provide only pathnames.

In Zmacs, c-m-Y has the effect of yanking from what's called the command history, a record of all editing commands which have used the mini-buffer. Immediately after a c-m-Y, m-Y has the usual effect.

One nice property of m-Y you might not immediately discover is that it takes numeric arguments, including negative ones. An argument of, say, 4, gives you the fourth next element in the history you're examining. An argument of -1 gives you the one you just m-Y'ed past; I usually use this one because because I was too quick on the m-Y trigger. For example, try typing this sequence to your Lisp Listener:

```
c-m-Y m-Y m-Y m-Y m-- m-Y
```

The ability to yank previous inputs into a Lisp Listener raises an interesting question: how does the input editor know when you're finished editing and ready for the input to be sent off to Lisp? Normally, if you just type your input without any yanking, the input editor knows you're done when you type some sort of delimiter at the end of the input string, like a close paren, to complete a well-formed Lisp expression. But if you've yanked an already well-formed expression, how can you complete it? The answer is that there is a special *activation character*. It is the End key. Pressing End while anywhere within a well-formed expression tells the input editor you're done, and it sends your input off to Lisp. So if you've yanked a previous input with c-m-Y, you can press End immediately to re-evaluate the same expression, or you can edit it some and press End when finished, to evaluate the modified expression.

The input editor also has commands which access online documentation of Lisp functions. These are:

- c-sh-A, which displays the argument list of the function whose name you have typed.

- m-sh-A, which displays the Symbolics documentation for the function whose name you have typed

- m-sh-D, which prompts you for the name of any Symbolics documentation topic and displays it on the screen.

So if I type "(with-open-file " to a Lisp Listener, and then press c-sh-A, the following will appear on my screen:

```
WITH-OPEN-FILE (MACRO): ((STREAM-VARIABLE FILENAME . OPTIONS)
                         &BODY BODY)
```

c-sh-A, m-sh-A and m-sh-D also work in Zmacs.

4.5 Mouse Sensitivity

With Genera 7.0, Symbolics has introduced a wholly new method of displaying output to and accepting input from the user. Every time anything interesting is displayed on most windows, the system remembers the object and its appearance on the screen. Advantages:

1. The Window System remembers what was on the window before the output was scrolled off the top. This means that you can *scroll back* to earlier parts of your output to see it again.

2. Everything on the screen is potentially *mouse sensitive*.

Let's take an example. Suppose you want to look at the contents of a file, but you don't remember what you called it. Here is what you might do:[10]

You type:

```
Show Directory LISP-LORE:EXAMPLES;CARD-GAME;*.LISP
```

and the system displays:

```
LISP-LORE:EXAMPLES;CARD-GAME;*.LISP.NEWEST
  375 free, 69595/69970 used (99%, 3 partitions)
       (LMFS records, 1 = 4544. 8-bit bytes)
    card-definitions.lisp.20    2    6509(8) !   10/15/86  ...
    card-places.lisp.31     3   11423(8) !    10/17/86 17:09:...
    card-presentation-types.lisp.7    2    6014(8) !    10/0...
    card-system.lisp.2    1     450(8)        10/02/86 18:49:3...
    card-table.lisp.21    4   14205(8) !    10/16/86 16:27:5...
    gaps-game.lisp.9    2    3866(8)  !    10/14/86 16:09:32...
    spider-game.lisp.13    2    3978(8)  !    10/15/86 17:43:...

16   blocks in the files listed
```

You could then type:

```
Edit File        and stop.
```

At this point, the `Edit File` command prompts you for the path-name of a file, and you could type one in. However, it might be quicker for you to use the mouse to point at the one you want and click on it.

---

[10]I have truncated the output lines to fit on the page. They don't actually appear with "..." on your screen. Yes, the columns really don't line up on your screen.

```
Edit File CD:>rsl>book>examples>card-game>card-system.lisp.2
```

Note that when I clicked on it, the entire pathname became part of my input, just as if I had typed it. If, for example, I decide that I really want to read version 1 of card-system.lisp, I can change the version number before I type Return.[11] If I click left on the pathname while holding the Shift modifier key, the CP behaves as if I had pressed the Return key after typing the pathname.

Now, the interesting thing about this mouse-sensitivity is that it is *input context sensitive*. For example, pathnames are not mouse-sensitive unless the machine is waiting for you to type a pathname. One refinement: if you are trying to read, say, a command, and you or the system have provided a *translation* which converts pathnames into commands, then pathnames will be mouse-sensitive when the system is trying to read a command. For example, if you click on a pathname when the command processor is trying to read a command, it is translated into the Show File command with that pathname as its argument.

In general, you can tell if an item displayed on the screen is mouse sensitive by pointing at it with the mouse. If it is, a box will be drawn around the sensitive output; this box will go away when the mouse is moved (or the object is no longer sensitive). At the bottom of the screen, there is a (usually) black area called the *Mouse Documentation* area. It will contain some text which describes what will happen if you click the mouse. The first line of its display will contain the results

---

[11] If the pathname I wanted to click on had disappeared off the top of the screen, I could press m-Scroll enough times until it reappeared, or click on the scroll bar in the left-hand margin. See the section "Looking Back Over Your Output (Scrolling)" in *User's Guide to Symbolics Computers.*

of clicking on the indicated object with the left, middle or right buttons. The second line will tell you what other combinations of shift keys (*e.g.*, control, meta, shift) will give you other options.

For example, let's point again to that pathname in the Show Directory output. The mouse documentation window will contain:[12]

```
Mouse-L: Show File (file) CD:>rsl>; Mouse-M: (DESCRIBE '#P...
To see other commands, press Shift, Control, Control-Shift...
```

As you can see, what happens when you click the mouse on an object depends on two things: what input context you're in, and what buttons you press. For example, if you're in the middle of typing in a Lisp expression, a pathname is sensitive as a Lisp object. An example:

You type:

```
Command: (fs:file-properties
```

and click on the pathname card-system.lisp, and then type a right parenthesis. Your screen might look like this:

```
Command: (fs:file-properties
          '#P"CD:>rsl>book>card-game>card-system.lisp.2")
(#P"CD:>rsl>book>examples>card-game>card-system.lisp.2"
  :GENERATION-RETENTION-COUNT 2 :LENGTH-IN-BLOCKS 1
  :REFERENCE-DATE 2739058608 :MODIFICATION-DATE 2737676975
  :CREATION-DATE 2737676975 :AUTHOR "rsl" :BYTE-SIZE 8
  :LENGTH-IN-BYTES 450 ...)
```

---

[12] edited slightly for readability

4.6 Poking Around

Many features of the programming environment will make your programming life much simpler. Learning these tools is often hit-or-miss, especially if there are no experienced Lisp Machine users at your site to suggest them to you. Here is a hodge-podge of commands and functions which will get you started in your exploration.

The function **who-calls** takes an argument, usually a symbol, and returns the list of functions which use that symbol as a function, a constant or a variable. It also returns declared variables which contain that object, either directly or as a list element. The editor commands List Callers and Edit Callers use this function to take you to the source of the calling functions.

In previous releases, **who-calls** was very slow. In Genera 7.0, it maintains a database of functions and their callers, and is thus much faster. By default, the database contains only those functions which you have defined in the current session (*i.e.*, since you cold booted), so the system software isn't searched. You or your site administrator can build a world which contains a full database containing the entire system. See the section "Enabling the Who-Calls Database At Your Site" in *Site* Operations.

The CP command Find Symbol searches for symbols whose name contains a string you provide. You can limit the search to only function names, flavor names, variable names, and so forth. You can also specify the packages to search.[13]

---

[13] You can also use the function **apropos** for this, but it's less flexible.

The command Show Compiled Code *disassembles* a function. This shows you the Lisp Machine instructions the compiler generates from your source code. The editor command Disassemble and the Lisp function **disassemble** do the same thing. You can disassemble flavor methods, including combined methods, by this method as well:

```
Show Compiled Code
        "(flavor:method :tyi si:interactive-stream)"
```

produces the following output:

```
Disassembled code for (FLAVOR:METHOD :TYI SI:INTERACTIVE-STREAM)
  0   ENTRY: 3 REQUIRED, 1 OPTIONAL
  1   PUSH-NIL
  2   PUSH-INDIRECT #'(DEFUN-IN-FLAVOR SI:TYI-INTERNAL SI: ...
  3   PUSH-LOCAL FP|0          ;SELF
  4   PUSH-LOCAL FP|1          ;SYS:SELF-MAPPING-TABLE
  5   PUSH-CONSTANT ':ANY-TYI
  6   PUSH-LOCAL FP|3          ;SI:EOF
  7   FUNCALL-4-RETURN
```

Finally, clicking left on a function name displayed by any disassembly will disassemble that function.

The Lisp Listener loop maintains certain variables whose values can be extraodinarily helpful. *, for instance, is always bound to the value returned by the last form you typed in; / is a list of all the values returned.[14] ** is the second-to-last value, and *** the third. Beyond that, you can click on old values, scrolling back up to them if necessary. See the section "Looking Back Over Your Output (Scrolling)" in *User's Guide to Symbolics Computers*.

---

[14] remember that Lisp forms can return more than one value.

4.7 Fun and Games

More definitions from *The Hacker's Dictionary* (Guy L. Steele Jr., *et al*), prompted by my spontaneous use of the term *loser*.

LOSE *verb.*

1. To fail. A program loses when it encounters an exceptional condition or fails to work in the expected manner.

2. To be exceptionally unaesthetic.

3. Of people, to be obnoxious or unusually stupid (as opposed to ignorant). See LOSER.

DESERVE TO LOSE *verb.* Said of someone who willfully does THE WRONG THING, or uses a feature known to be MARGINAL. What is meant is that one deserves the consequences of one's losing actions. "Boy, anyone who tries to use UNIX deserves to lose!"

LOSE, LOSE *interjection.* A reply or comment on an undesirable situation. Example: "I accidentally deleted all my files!" "Lose, lose."

LOSER *noun.* An unexpectedly bad situation, program, programmer, or person. Someone who habitually loses (even winners can lose occasionally). Someone who knows not and knows not that he knows not. Emphatic forms are "real loser," "total loser," and "complete loser."

LOSS *noun.* Something (but not a person) that loses: a situation in which something is losing.

WHAT A LOSS! *interjection.* A remark to the effect that a situation is bad. Example: Suppose someone said, "Fred decided to write his program in ADA instead of LISP."

The reply "What a loss!" comments that the choice was bad, or that it will result in an undesirable situation – but may also implicitly recognize that Fred was forced to make that decision because of outside influences. On the other hand, the reply "What a loser!" is a more general remark about Fred himself, and implies that bad consequences will be entirely his fault.

LOSSAGE *(lawss':j) noun.* The stuff of which losses are made. This is a collective noun. "What a loss!" and "What lossage!" are nearly synonymous remarks.

WIN

1. *verb.* To succeed. A program wins if no unexpected conditions arise. Antonym: LOSE.

2. *noun.* Success, or a specific instance thereof. A pleasing outcome. A FEATURE. Emphatic forms: MOBY win, super-win, hyper-win. For some reason "suitable win" is also common at MIT, usually in reference to a satisfactory solution to a problem. Antonym: LOSS.

BIG WIN *noun.* The results of serendipity.

WIN BIG *verb.* To experience serendipity. "I went shopping and won big; there was a two-for-one sale."

WINNER *noun.* An unexpectedly good situation, program, programmer, or person. Albert Einstein was a winner. Antonym: LOSER.

REAL WINNER *noun.* This term is often used sarcastically, but is also used as high praise.

WINNAGE *(win':j) noun.* The situation when a LOSSAGE is corrected or when something is winning. Quite rare. Usage: also quite rare.

WINNITUDE *(win':-tood) noun.* The quality of winning (as opposed to WINNAGE, which is the result of winning).

4.8 Problem Set

Questions

1. Define a function which produces a concordance of all symbols in a package. That is, it makes two alphabetical lists of all the functions and all the variables in the package. The list of functions should be a list of lists; the first element should be the name of the function and the rest of the list a list of its callers. [Hint: use **do-symbols** or **do-local-symbols**.]

2. Run that function in the background. In order to be useful, it should save the lists someplace you can get to them. Notify the user when the function is done.

3. Make a function which displays a symbol concordance on a pop-up window. Hook it up to a Function key, like, say, Function Square, using **tv:add-function-key**. If you're unsure how to do this, look at the source for the function **tv:kbd-finger** with the Zmacs command m-..

Answers

1. Here is one way to do it:

```
(defun make-symbol-concordance
       (&optional (package *package*))
  (let ((functions) (variables))
    (do-local-symbols (symbol package)
      (when (boundp symbol)
        (push symbol variables))
      (when (fboundp symbol)
        (push (cons symbol (who-calls symbol))
              functions)))
    (values variables functions)))
```

Using **do-local-symbols** avoids all the symbols in the Lisp package, like **car** and +; these functions have *many* callers, most of which are uninteresting for this use.

2. You should use the function **tv:notify** to tell the user when you're done doing something in the background.

```
(defvar *variables*)
(defvar *functions*)

(defun make-symbol-concordance-background
       (&optional (package *package*))
  (process-run-function (:name "Make Concordance"
                         :priority -1)
    (lambda ()
      (let ((*standard-output* #'sys:null-stream))
        (multiple-value-setq (*variables* *functions*)
          (make-symbol-concordance package)))
      (tv:notify nil "Finished making concordance."))))
```

I bound ***standard-output*** to the "null stream" because
who-calls does some output on that stream.

3. I chose to make this answer a little more complicated, so
you can see how to deal with a numeric argument to a
Function key.

```
(defun show-symbol-concordance-function-key
       (numeric-arg &aux (package *package*))
  (tv:with-pop-up-window "Concordance window"
    (when numeric-arg
      (setq package (accept 'package
                            :prompt "Package")))
    (multiple-value-bind (vars funs)
        (let ((*standard-output* #'sys:null-stream))
          (make-symbol-concordance package))
      ;; Show the variables
      (format t "Variables in package ~A:" package)
      (dolist (sym vars) (format t "~%~A" sym))
      ;; Show the functions.
      (format t "~2%Functions in package ~A:" package)
      (dolist (fun funs) (format t "~%~A" (car fun))
        (when (cdr fun)
          (format t " called by: ~{~S~^, ~}"
                  (cdr fun))))
      ;; Allow the user to say when she is done.
      (tv:type-a-space-to-flush *terminal-io*))))

;;; Hook it up to the Function key.
(tv:add-function-key
  #\Square
  'show-symbol-concordance-function-key
  "Show symbol concordance for a package.")
```

5. What's a Flavor?

(For a more detailed presentation of this material: See the section "Flavors" in *Symbolics Common Lisp: Language Concepts.* I have skipped many features of flavors which you may find useful and which are fully described there.)

The Flavor System[1] is the Lisp Machine's mechanism for defining and creating active objects, that is, objects which "remember" their state and "know" how to perform certain operations. A *flavor* is a class of such objects. Each such object is an *instance* of that flavor.[2]

Programming with flavors is especially useful when:

- You are modeling objects whose behavior might change over the course of time. For example, a veterinarian's as-

---

[1] See hacker's definition at end of chapter.

[2] The flavor system is similar to **defstruct**, in that they both provide a mechanism for structuring data. However, the Flavor System also provides a way to specify behavior of objects which depends on their type (flavor). See the macro **defstruct** in *Symbolics Common Lisp: Language Concepts.*

sistant program might want to model the eating behavior of dogs. Presumably, what they eat varies with their age and health.

- You are providing a generic interface to different kinds of objects. For example, a program which displays mailing lists should not be required to know whether its output is being displayed on a printer or a screen; it should use an "output stream" object which always defines the operation of "write this character at the current position."

There are two primary characteristics of a flavor:

1. The set of state variables an instance of that flavor has. These variables are called *instance variables*.

2. The set of operations which may be performed on all instances of that flavor. These operations are implemented by functions called *methods*.

5.1 Instance Variables

Every instance of a given flavor has the same set of instance variables. The *values* of those variables are likely to be different from one instance to another.

When an instance is created, the instance variables are all initialized. Those initial values can be declared at compile time or supplied at run time. If no value is specified, the instance variable is left uninitialized, and the Lisp Machine will signal an uninitialized-variable error if the variable is referenced before it is set.

Flavors are defined with the **defflavor** special form. Here is a simple definition of a flavor named **ship**, which might be used in a program for an outer space game:

```
(defflavor ship
        (x-position y-position
         x-velocity y-velocity
         mass)
        ())
```

It states that all instances of the flavor **ship** will have five instance variables, as listed. (The empty list following the instance variables is related to a feature we'll consider in a minute). Of course, two different **ships** will have different places to store each of these variables, but it always makes sense to ask, for example, the value of the instance variable **x-velocity** of any given **ship**.

5.2 Methods

A *method* is a function which provides behavior for instances of a flavor. All instances of a given flavor have the same methods.

Methods are defined with **defmethod**, which looks very much like **defun**. Using **defmethod**, the programmer specifies a method name, an argument list, and a body. The body will be executed in an environment in which the names of the instance variables will refer to the instance variables of the specific instance.

Here are two methods for the **ship** flavor, to provide the generic operations **speed** and **direction**:

```
(defmethod (speed ship) ()
  (sqrt (+ (expt x-velocity 2) (expt y-velocity 2))))
```

```
(defmethod (direction ship) ()
  (atan y-velocity x-velocity))
```

Instance variables are *lexically scoped* names within the body of methods. The way you refer to an instance variable inside the body of a method is by using its name. If you need to change the value of an instance variable, you can do so with **setf** or **setq**.

We might also wish to have methods which allow one to read or modify the values of a **ship**'s instance variables from someplace other than within a method. For example:

```
(defmethod (ship-x-position ship) ()
  x-position)
```

Writing one of these methods for every instance variable would be tedious and error-prone. Fortunately, there is an option to **defflavor** which automatically generates accessors (like **defstruct** accessors) for any instance variables you choose, including all of them. There is also an option which causes defflavor automatically to generate **setf** methods for those accessors.

```
(defflavor ship
           (x-position y-position
            x-velocity y-velocity
            mass)
           ()
         :readable-instance-variables
         (:writable-instance-variables x-position y-position
                                       x-velocity y-velocity))
```

This creates such accessors as **ship-x-position** and **ship-mass**. You can use **setf** on the position and velocity instance variable accessors, but not **mass.**[3]

5.3 Making Instances

To make an instance of a flavor, we use the **make-instance** function:[4]

```
(setq my-ship (make-instance 'ship))
```

This will return an object whose printed representation looks like #<SHIP 255645543>. (The funny number will be the virtual memory address, in octal, of the instance.)

To "call" a method of an instance, you merely use it as a function. We can now do things like:

```
(setf (ship-x-velocity my-ship) 1000)   => 1000
(setf (ship-y-velocity my-ship) 500)    => 500
(speed my-ship)                         => 1118.0339
```

In addition to the instance variables, another very important variable is available to the body of a method. The value of the variable **self** will be the instance itself. **self** is often used to ask the object to perform another operation:

---

[3]**:readable-** and **:writable-instance-variables** may either be present alone, meaning they apply to all the instance variables, or at the head of a list, meaning that they apply only to the listed variables. This allows the programmer complete control over modularity. If it doesn't make sense to allow programmers to read or change a part of your instance, you don't have to permit them to do so.

[4]Windows are a special kind of flavor. To make instances of windows, you should use **tv:make-window** instead of **make-instance.**

```
(defmethod (check-speed ship) ()
  (when (> (speed self) 3.0e8)
    (error "travel at rates greater than the ~
                speed of light is not permitted")))
```

5.4 Initial Values for Instance Variables

Instances of our **ship** flavor start out with all their instance variables unbound. Using the **ship-x-position** function on one, for instance, would result in an unbound-variable error. But there are two ways to arrange for initial values to be assigned to an instance when it is made. If you have used the **:initable-instance-variables** option to defflavor, then you may specify the initial values in the call to **make-instance**. So, with this **defflavor**:

```
(defflavor ship
        (x-position y-position
         x-velocity y-velocity
         mass)
        ()
      :readable-instance-variables
      (:writable-instance-variables x-position y-position
                                    x-velocity y-velocity)
      :initable-instance-variables)
```

you could use this call to **make-instance**:

```
(make-instance 'ship :x-position 30 :y-position -150
            :mass 10)
```

The instance variables named in the call will have the specified initial values. Instance variables not mentioned will be unbound, as before. Now, suppose you want all instances to have certain initial values for certain instance variables. Perhaps you want the **x-velocity** and **y-velocity** of *all* new **ships** to be 0. You could specify so in every call to **make-instance**. But there is an easier way. You can specify in the defflavor what initial value you wish the instance variables to have. Here's our next version of the **defflavor** for **ship**:

```
(defflavor ship
        (x-position
         y-position
         (x-velocity 0)
         (y-velocity 0)
         mass)
        ()
     :readable-instance-variables
     (:writable-instance-variables x-position y-position
                                   x-velocity y-velocity)
     :initable-instance-variables)
```

Now all **ships** will start out with x- and y- velocities of 0 – unless you specify otherwise in the **make-instance**. As before, the position and mass will be unbound by default. An initial value specified in **make-instance** will override any default initial values given in the defflavor.

Here is a slightly more complex example, taken from the flavor documentation:

```
(defvar *default-x-velocity* 2.0)
(defvar *default-y-velocity* 3.0)
```

```
(defflavor ship
        ((x-position 0.0)
         (y-position 0.0)
         (x-velocity *default-x-velocity*)
         (y-velocity *default-y-velocity*)
         mass)
        ()
    :readable-instance-variables
    (:writable-instance-variables x-position y-position
                                  x-velocity y-velocity)
    :initable-instance-variables)
```

```
(setq another-ship (make-instance 'ship :x-position 3.4))
```

What will the values of the new **ship**'s instance variables be?[5] The function **describe** can be useful for seeing what your instance has in its instance variables. In general, **describe** tries to print helpful information about its argument. When applied to an instance, it prints all the instance variables. For example,

```
(describe another-ship)    would print
```

```
#<SHIP 410010274>, an object of flavor SHIP,
  has instance variable values:
    X-POSITION:           3.4
    Y-POSITION:           0.0
    X-VELOCITY:           2.0
    Y-VELOCITY:           3.0
    MASS:                 unbound
```

---

[5]Answer: 3.4 for **x-position** (the **make-instance** specification overrides the default of 0.0), 0.0 for **y-position** (the default), 2.0 for **x-velocity** and 3.0 for **y-velocity** (the values of the two global variables); **mass** will be unbound.

5.5 Methods for Make-instance

One method which you may wish to write is for the **make-instance** generic function. Each time you make an instance, the last thing that is done before the **make-instance** returns the instance to you is that it calls the **make-instance** method for your new instance. For example:[6]

```
(defmethod (make-instance ship)
              (&key name &allow-other-keys)
    (push self *all-the-ships*)
    (setf (gethash *ships-registry* name) self))
```

To allow keywords other than instance variable names for use with **make-instance**, use the **defflavor** option **:init-keywords**. See the special form **defflavor** in *Symbolics Common Lisp: Language Concepts*.

5.6 Mixing Flavors

The real power of Flavors lies in its facility for producing new flavors by combining existing ones. Suppose we wished to add asteroids to our game. In many ways, an asteroid behaves the same way a ship does: it obeys Newton's laws, special relativity, and so forth. In fact, all of our current **ship**'s instance variables and methods would be appropriate. But we do want to have two distinct kinds of object, because as our program becomes more complete, **ships** and **asteroids** will not behave the same. **ship**, for instance, might use an instance variable for its

---

[6]You should always specify **&allow-other-keys** for **make-instance** methods because all the initialization keywords passed to **make-instance** are passed to your method, whether you care about them or not.

engine power, or we might want to give each **ship** a name. And we might want to characterize each asteroid's composition (perhaps part of the game requires replenishing resources by mining asteroids).

One way to handle the situation would be to duplicate the Lisp code for the common functionality in both flavors. Such duplication would clearly be wasteful, and the program would become far more difficult to maintain – any modifications would have to be repeated in both places. A better approach would be to isolate the common functionality and make it a flavor in itself. We can call it **moving-object**. Now the **ship** and asteroid flavors can be built on **moving-object**. We just need to specify the added functionality each has beyond that provided by **moving-object**. The **defflavor** for **moving-object** can be exactly like our existing **defflavor** for **ship**. The new **ship** defflavor will have **moving-object** specified in its list of component flavors, which up until now has been an empty list.

```
(defflavor ship (engine-power name)
           (moving-object)
  :readable-instance-variables
  :initable-instance-variables)
```

And asteroid:

```
(defflavor asteroid (percent-iron)
           (moving-object)
  :readable-instance-variables
  :initable-instance-variables)
```

ship and **asteroid** both inherit all of **moving-object**'s instance variables (including their default values) and all of its methods. They are each *specializations* of the abstract type **moving-object**. And the specialization could continue. We could define a **ship-with-passengers** flavor, built on ship, with an added instance variable **passengers**, and added methods for **add-passenger** and **remove-passenger**.

Figure 2. Flavor inheritance hierarchy for **si:indirect-escape-input-stream**.

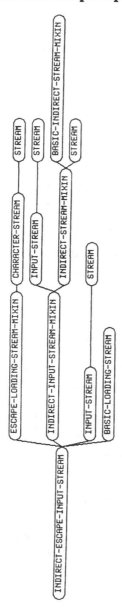

A flavor is not limited to having only one component flavor – it may have any number. So the set of components for a given flavor is actually a tree, consisting of all the flavor's direct components, and all of their direct components, and so on. Figure 2 shows the tree for flavor **si:indirect-escape-input-stream**, a flavor of file stream.

5.7 Combined Methods

Simply saying that a flavor inherits all the methods of its components sweeps an important issue under the rug. What happens if more than one of its components define methods for the same generic function? Which gets used?

It depends on the ordering of the component flavors. The Symbolics documentation describes this pretty well: See the section "Mixing Flavors" in *Symbolics Common Lisp: Language* Concepts. Three basic rules govern the ordering in most flavor combinations:

1. A flavor always precedes its own components.

2. The local ordering of flavor components is preserved, *i.e.,* the order in which they appear in the **defflavor** form.

3. Duplicate flavors are eliminated from the ordering: if a flavor appears more than once, it is placed as close to the beginning of the ordering as possible, while still obeying the rules.

If more than one component flavor defines a method for a given generic function, with the kind of methods we have seen so far, the one which appears first on the list is taken as the combined flavor's method for that generic function. In particular, this means that any methods (again, of the type we have seen so far) defined locally in the new flavor will supersede all methods (for the same generic function) defined in any of its component flavors, since the new flavor is first on the ordered list.

For example, a **moving-object** might have a method for the generic function **accelerate**, which did nothing. A **ship**, or perhaps the **engine-mixin** flavor on which it is built, might have an **accelerate** method which calculated the new velocities based on the mass of the ship and how much thrust its engine could generate:[7]

```
(defmethod (accelerate moving-object) (pct-thrust delta-time)
  ;; One way to ignore arguments.
  (ignore pct-thrust delta-time))

(defflavor engine-mixin
        (thrust
         orientation)
        (moving-object)
   :initable-instance-variables)

(defmethod (accelerate engine-mixin) (pct-thrust delta-time)
  (let* ((true-thrust (* thrust pct-thrust .01))
         (thrust-x (* true-thrust (cos orientation)))
         (thrust-y (* true-thrust (sin orientation))))
    (incf x-velocity (* thrust-x delta-time))
    (incf y-velocity (* thrust-y delta-time))))

;;; Redefine ship flavor to use new component flavors
(defflavor ship
```

---

[7]Of course, an asteroid, built on **moving-object,** would inherit the do-nothing method, unless it also had an engine mixed in.

```
      ()
      (engine-mixin moving-object))
```

As I've hinted, there are more kinds of methods than we have
so far seen. All our methods have been what are called
"primary" methods, and by default, when there is more than
one primary method for the same generic function in the ordered list of component flavors, the one which appears first
overrides all others. But sometimes you don't want to completely override the inherited primary method; sometimes you
would like to specify something to be done in addition to the
action of the inherited method rather than instead of. Then
you would define a :before or an :after method, often called
before and *after daemons*.

Here's how it works. Suppose we add the following defmethod:

```
(defmethod (accelerate moving-object :after)
           (ignore delta-time)
           ;; Another way to ignore arguments
  (incf x-position (* x-velocity delta-time))
  (incf y-position (* y-velocity delta-time)))
```

moving-object already has a primary **accelerate** method. Once
this new **accelerate :after** method is defined, both **asteroid** and
ship will have a "combined method" for **accelerate**, consisting
of a call to the appropriate primary method followed by a call
to the **moving-object :after** method. Any number of flavors in
the ordered list of components may provide daemons. They will
all be included in the resulting combined method. The primary
method which appears first in the list will be called after all
the before daemons (even if some of the before daemons appear
later in the list than the primary method) and before all the
after daemons. The **:before** daemons will be executed in the
order in which they appear in the flavor ordering; the **:after**
daemons will be executed in the opposite order.

The value returned by a combined method is exactly the value returned by the primary method – before and after daemons are executed only for side effect, *i.e.*, their return values are ignored. It is allowable to have before and after daemons for a generic function which has no primary method; in such a case the combined method will return **nil**.

Unlike all the other methods we have described so far, **make-instance** "primary" methods do *not* override each other. For example:

```
(defmethod (make-instance ship)
           (&key name &allow-other-keys)
  (push self *all-the-ships*)
  (setf (gethash *ships-registry* name) self))

(defmethod (make-instance engine-mixin)
           (&key fuel-source &allow-other-keys)
  (if fuel-source
     (setq fuel (obtain-fuel fuel-source fuel-capacity))
     (setq fuel nil)))
```

The combined method for **ship** will both register the ship and fill the fuel tank.

5.8 Whoppers

Before and after daemons provide a lot of flexibility (perhaps more than you'd like to have just yet), but sometimes not enough. Frequently, a situation demands altering the context in which a primary method runs. Typical cases include:

- binding a special variable to some value around the execution of the primary method

- putting the primary method into an **unwind-protect** or inside a **catch**.[8]

- deciding in some cases to skip the primary method altogether, or call it more than once.

- modifying the argument list before invoking the method.

The kind of method which can do all of these is called a whopper. Whoppers are best explained by example. Here are several, which handle the cases I just listed. To understand them, you'll need to know that **continue-whopper** is a system-provided function which calls the regular (non-whopper) methods (also called the *continuation*) for this generic function.

```
(defwhopper (some-generic some-flavor) (arg1 arg2)
  (let ((*some-special-variable* (compute-value arg1)))
    (continue-whopper arg1 arg2)))

(defwhopper (operate-reactor reactor) (arg1 arg2)
  (unwind-protect
      (progn (slide-out-control-rods self)
             (continue-whopper arg1 arg2))
    (slide-in-control-rods self)))
```

Unlike before and after daemons, whoppers have control over the value returned by the combined method. They most commonly just pass up the value(s) returned by **continue-whopper** (which will be whatever the primary method returns, as before), but they needn't. I could, for instance, do this:

---

[8]If **unwind-protect** or **catch** are unfamiliar, you might want to look back to chapter 3.

```
(defwhopper (calculate doubling-mixin) (arg1 arg2)
  (* 2 (continue-whopper arg1 arg2)))
```

And since continue-whopper is just a function like any other, there's no reason you couldn't do something like this:

```
(defwhopper (some-generic yet-another-doubling-mixin)
            (arg1 arg2)
  (continue-whopper arg1 (continue-whopper arg1 arg2)))
```

Or this:

```
(defwhopper (some-generic some-flavor) (arg1 arg2)
  (when (some-special-test arg1)
    (list (continue-whopper arg1 arg2)
          (continue-whopper nil arg2)))))
```

Note that the last example calls the continuation either twice or not at all; the second time it calls it, **arg1** is **nil**, not what the user passed in. For a real example in the system source, try looking at the whopper for the method **:tyo** for the flavor **si:ascii-translating-output-stream-mixin**.[9] One point about ordering needs to be clarified. A whopper surrounds not just the primary method, but all the before and after daemons, too. So suppose flavor **out** is built on top of **in**, and both **out** and **in** have a whopper, a before daemon, an after daemon, and a primary method for the generic function **mumble**. **out**'s combined method for **mumble** would look like Figure 3.

In some older code, you may see a similar construct called a *wrapper* (defined, of course, with **defwrapper**.) This was the predecessor of the whopper, but now that the whopper exists,

---

[9]To see the source, use the m-. editor command on the definition (flavor:whopper :tyo si:ascii-translating-output-stream-mixin).

out-before

in-before

out-whopper   in-whopper   out-primary

in-after

out-after

Figure 3.   Structure of combined method

there is seldom any need to use wrappers. Wrappers are much more difficult to write and debug, although they can produce slightly faster methods, because they are macros instead of functions.   To obtain the benefits of wrapper efficiency while using whopper-like syntax: See the macro **defwhopper-subst** in *Symbolics Common Lisp: Language Dictionary*.

The Flavor System defines other ways to combine methods I have not described.   For example, rather than returning only the value of the primary method (or the whopper), you might want to return the sum of all the values (the **:sum** method combination type) or choose only one of the methods based on another argument to the generic function (the **:case** method combination type).   This is well-documented by Symbolics. See the section "Method Combination" in *Symbolics Common Lisp: Language Concepts*.

5.9 Internal Interfaces

One of the reasons to use Flavors is to expose only a well-defined interface to your implementation of a type of object. If you define all the functions which know about your objects as generic functions, however, other users will think that these methods are part of the interface, and will call them. If you ever want to change how your flavor is implemented, it will be more difficult to maintain compatibility.

There is another way to define a function which has access to all the instance variables (and **self**) of a flavor. The special form **defun-in-flavor** creates a function which can be called *only* from a method or another function in that flavor. These functions are *lexically scoped* within the body of each method and function of the flavor.

5.10 Vanilla Flavor

flavor:vanilla[10] is the flavor on which all other flavors are built. Even if your **defflavor** specifies no components, your flavor will still have vanilla flavor mixed in, because the flavor system does it automatically.

Don't complain. Vanilla flavor is very handy. It provides several extremely important methods. The **sys:print-self** method is called whenever an instance is to be printed. (The representation of the first **ship** instance we made, #<SHIP 255645543>, was actually printed on my monitor by **ship's sys:print-self** method, inherited from vanilla flavor. The

---

[10]See hacker's definition at end of chapter.

method **:describe**, used by the function **describe**, prints the type of object and its instance variables.

Vanilla flavor also supplies other methods. See the section "Generic Functions and Messages Supported by **flavor:vanilla**" in *Symbolics Common Lisp: Language Concepts*.

5.11 The Flavor Examiner Tools

The power of Flavors does not come without its costs, namely potentially increased complexity. The system provides a number of tools to help disentangle the interrelationships of the various parts of the Flavors database.

There are three places to find these tools. In the Flavor Examiner window, they are all assembled as a menu and a set of display panes. To get to the Flavor Examiner window, type Select X, or use the system menu or the Select Activity command.

In the command processor, all the flavor examination commands begin with "Show Flavor ...": If you type:

```
Show Flavor
```

and press the Help key, your output might look like:

```
These are the command names starting with "Show Flavor":
Show Flavor Components      Show Flavor Initializations
Show Flavor Dependents      Show Flavor Instance Variables
Show Flavor Differences     Show Flavor Methods
Show Flavor Functions       Show Flavor Operations
Show Flavor Handler
```

These commands are also available in the editor, by using meta-X.

The other Flavors-related command you might wish to use is Show Generic Function. All of these commands are documented in the Symbolics documentation. See the section "Flavors Tools" in *Symbolics Common Lisp: Language Concepts*.

One other useful tool in the editor deserves mention here. The meta-X command Show Effect Of Definition will show you what changes in your world a given form in your editor buffer will cause, if you were to compile it using c-sh-C. This is not only useful for flavors, but also methods, whoppers, and so forth.

5.12 Message Passing

In earlier (*i.e.*, pre-release 7.0) versions of the system, Flavors, although conceptually similar, used a different syntax than it does today. The syntax for method definition was slightly different, and the syntax for method invocation was pretty baroque. This earlier syntax was called *message passing*, although what was meant by message passing is not what you might intuitively think (especially if you're familiar with the use of that term in the computer science literature).

In genuine message passing, sending a message to an object causes a (potentially) asynchronous response to your message. This is just like the way, say, an office works: you "pass a message" to your secretary, or your boss, or your colleague, and (presumably) they do something for you. Meanwhile, you're free to go off and do other things.

In Flavors "message passing", the syntax often suggested that this asynchronous response would happen. In fact, all sending a message did in Flavors was to look up the appropriate method and invoke it (synchronously), just like generic functions in the current release. Thus, the old syntax was somewhat misleading.

In addition, the old syntax made certain "Lisp-y" things hard to do. For example, a common programming technique in Lisp is to pass functions around as arguments to other functions. With the current syntax, you can use a generic function anywhere you could use any other function. In the old syntax, this was harder.

The only reason you care about this now is that a lot of software was written using the old style of flavors. This code still works in the current implementation, and will continue to work for the forseeable future. A lot of the *system* software is still written using message-passing style, including the window system. See the section "The Window System," page 120.

Here are some examples of "old-style" and "new-style" flavors code.

Old style:
```
(defmethod (hollerith-stream :tyi) ()
  (convert-punch-to-ebcdic (get-next-input-byte self)))
```

New style:
```
(defmethod (:tyi hollerith-stream) ()
  (convert-punch-to-ebcdic (get-next-input-byte self)))
```
or, better:
```
(defmethod (get-ebcdic-byte hollerith-stream) ()
  (convert-punch-to-ebcdic (get-next-input-byte self)))
```

Old style:
```
(defun get-next-byte (hollerith-stream)
  (ebcdic-to-char (send hollerith-stream :tyi)))
```

New style:

```
(defun get-next-byte (hollerith-stream)
  (ebcdic-to-char (get-ebcdic-byte hollerith-stream)))
```

Old style:
```
(defflavor massive-object
        (mass)
        ()
  :settable-instance-variables
  :gettable-instance-variables)

(defun do-something-to-massive-object (object)
  (let ((mass (send object :mass)))
    ...
    (send object :set-mass new-mass)))
```

New style:
```
(defflavor massive-object
        (mass)
        ()
  :readable-instance-variables
  :writable-instance-variables)

(defun do-something-to-massive-object (object)
  (let ((mass (massive-object-mass object)))
    ...
    (setf (massive-object-mass object) new-mass)))
```

5.13 The Window System

Much of the "old-style" Flavors code in Genera 7.0 is in the *window system*. The window system implementation is older than the current implementation of Flavors. In prior releases, programmers who wanted to do anything fancy at all with the user interface pretty much *had* to define their own flavors of windows. Since documentation was often scanty, many of these window flavors are highly convoluted, are often hard to understand, and frequently use the window system at the wrong level of modularity.

The current release of the system defines a much simpler programmer interface to the window system, and most likely you won't have to use flavors to write new window-using software. Unfortunately, a lot of software has already been written which does things the old way, defining flavors of windows and various methods on them. To aid your comprehension of these programs, in case you need to read or maintain one of them, here are some hints.

1. Message passing – The entire window system is implemented using message-passing.

2. The :init method – In the earlier version of Flavors, the way to make something happen when you made an instance was by defining an :init method. The current flavor system still **sends** the :init message to newly-created instances if they handle that message. :init is always sent *after* the **make-instance** method is run.

   Certain flavors require that you write methods which run after their :init methods. If you tried writing methods to run after **make-instance** instead, it would not have the desired effect.

3. **:or** method combination – The **:mouse-click** method is combined using **:or** method combination. What this means is that the methods are run one at a time, until one returns something non-**nil**. If you try to read a **:mouse-click** method, you won't understand what's happening until you understand this concept.

5.14 Fun and Games

And from *The Hacker's Dictionary*, Guy L. Steele, Jr., *et al*:

FLAVOR *noun.*

1. Variety, type, kind. "Emacs commands come in two flavors: single-character and named." "These lights come in two flavors: big red ones and small green ones." See VANILLA.

2. The attribute that causes something to be FLAVORFUL. Usually used in the phrase "yields additional flavor." Example: "This feature yields additional flavor by allowing one to print text either right-side-up or upside down."

VANILLA *adjective.* Standard, usual, of ordinary FLAVOR. "It's just a vanilla terminal; it doesn't have any interesting FEATURES." When used of food, this term very often does not mean that the food is flavored with vanilla extract! For example, "vanilla-flavored wonton soup" (or simply "vanilla wonton soup") means ordinary wonton soup, as opposed to hot-and-sour wonton soup.

This word differs from CANONICAL in that the latter means "the thing you always use (or the way you always do it) unless you have some strong reason to do otherwise," whereas "vanilla" simply means "ordinary." For example, when MIT hackers go to Colleen's Chinese Cuisine, hot-and-sour wonton soup is the canonical wonton soup to get (because that is what most of them usually order) even though it isn't the vanilla wonton soup.

5.15 Problem Set

Questions

Part I

Let's write a geometry system.

1. Define a flavor for rectangles. A rectangle should remember its size. Define a generic function **draw-self** that takes a window and x and y coordinates as arguments and draws the rectangle. Test it out.

2. Define a mixin that draws itself on the screen when it is first created, and then re-draws itself when it is moved. This mixin should remember the location of the rectangle. Mix that flavor into your rectangle flavor.

3. Create a **move-self** method for your mixin flavor. It should update the location on the window.

4. Define a mixin that pushes each new instance of itself onto a list which is kept in a global variable. Mix that into your rectangle flavor.

5. Define some more flavors for other shapes.

Part II

The Flavor Examiner tools are quite powerful for helping debug flavor problems, but there are a few tools missing. Here is a tool you might write for practice: A Find Methods command, which prints the methods of an instance whose name matches a given string.

Hints

Part I

1. Use **graphics:draw-rectangle**. You probably won't see it if you test it on, say, the default Lisp Listener. Try using **dw:with-own-coordinates**.

2. It will probably have to remember the window on which it is drawn. Mix it into your rectangle shape, and see what happens when you create rectangles.

3. Consider the use of **tv:alu-xor** for both drawing and erasing.

4. Remember that **make-instance** methods don't overwrite each other.

5. You might want to abstract the draw/erase and remember-each-instance flavor into a higher-level flavor.

Part II

You can get a list of all the names of generic functions defined for an instance by sending the instance the **:which-operations** message. Remember (or find out) what the methods for setting writable instance variables are called.

To define a command, use **cp:define-command**. The type of objects you want to read are a string and a form to evaluate to give the object. You might want the form to be *, *i.e.*, the last value typed out, by default.

Answers

Part I

1. I named the flavor **rectangle**

```
(defflavor rectangle
         (width
          height)
      ()
   :initable-instance-variables
   (:required-init-keywords :width :height))

(defmethod (draw-self rectangle) (xpos ypos window)
   (let ((half-width (floor width 2))
         (half-height (floor height 2)))
      (graphics:draw-rectangle (- xpos half-width)
                               (- ypos half-height)
                               (+ xpos half-width)
                               (+ ypos half-height)
                               :stream window)))

(defun test-draw-self
       (rectangle xpos ypos
        &optional (window *standard-output*))
   (dw:with-own-coordinates (window)
      (draw-self rectangle xpos ypos window)))
```

2. Note that you have to know the location and window when you create the instance. I have made them required init keywords. If you really care, you should test to make sure they're really what they're supposed to be in the **make-instance** method.

```
(defflavor place-remembering-mixin
         (x-position y-position window)
```

```
      ()
   (:required-methods draw-self)
   :initable-instance-variables
   (:required-init-keywords :x-position :y-position
                            :window))

(defmethod (make-instance place-remembering-mixin)
           (&rest ignore)
  (draw-self self x-position y-position window))

(defflavor rectangle
        (width
         height)
        (place-remembering-mixin)   ; Added mixin
   :initable-instance-variables
   (:required-init-keywords :width :height))
```

3. It is better to write a **move-self** method than to make **xpos** and **ypos** writable because this allows for better modularity.

```
(defmethod (move-self place-remembering-mixin)
           (new-x new-y)
  (setf x-position new-x y-position new-y))

(defwhopper (move-self place-remembering-mixin)
            (new-x new-y)
  (draw-self self x-position y-position window)
  (continue-whopper new-x new-y)
  (draw-self self x-position y-position window))

(defmethod (draw-self rectangle) (xpos ypos window)
  (let ((half-width (floor width 2))
        (half-height (floor height 2)))
    (graphics:draw-rectangle (- xpos half-width)
```

```
(- ypos half-height)
(+ xpos half-width)
(+ ypos half-height)
:stream window
;; Draw using XOR
:alu tv:alu-xor)))
```

Putting the erase and draw steps in the whopper instead of in the method is unnecessary in this particular example, but consider the case where drawing and place-remembering are in separate mixins.

4. First, define a global variable for the list. Then, the mixin, its **make-instance** method, and finally mix it in.

```
(defvar *remembered-shapes* nil)

(defflavor shape-remembering-mixin
        ()
        ())

(defmethod (make-instance shape-remembering-mixin)
           (&rest ignore)
  (push self *remembered-shapes*))

(defflavor rectangle
        (width
         height)
        (shape-remembering-mixin
         place-remembering-mixin)
  :initable-instance-variables
  (:required-init-keywords :width :height))
```

5. Here's the mixin, and a **circle** flavor:

```
(defflavor basic-shape
        ()
        (shape-remembering-mixin
         place-remembering-mixin)
      :abstract-flavor)

(defflavor circle
        (radius)
        (basic-shape)
      :initable-instance-variables
      (:required-init-keywords :radius))

(defmethod (draw-self circle) (xpos ypos window)
      (graphics:draw-circle xpos ypos radius
                            :stream window
                            :alu tv:alu-xor))
```

Part II

Here is the command Find Method as defined in my init file. A couple of extra features in my version:

- The function names are presented as generic function names, rather than just printed as symbols.

- The argument list of the method is printed.

```
(cp:define-command (com-find-method :command-table "global")
    ((substring 'string
                :prompt "substring")
     (instance 'sys:expression
                :prompt "for value"
                :default '*))
    (loop for (nil . method) in
          (sort
              (loop for method-name in
```

```
                    (send (eval instance) :which-operations)
                as string-method-name =
                    (if (symbolp method-name)
                        method-name
                        (format nil "~S ~S"
                                    (second method-name)
                                    (first method-name)))
                when (string-search substring
                                        string-method-name)
                  collect (cons string-method-name
                                    method-name))
        #'string-lessp
        :key #'car)
    do
(dw:with-output-as-presentation
  (:object method :type 'sys:generic-function-name)
  (format t "~%~S: ~:A" method
    (arglist (or
                (si:function-spec-get
                  method 'flavor:generic)
                (si:function-spec-get
                  method
                  'flavor::compatible-generic)))))))))
```

6. User Interface

In this chapter I will talk about a part of the operating system environment of Genera, namely programming the user interface.

In previous versions of the Lisp Machine system, user interface programming has often proven to be the largest part of people's systems. The part of their program which was actually solving their problem was often smaller than the part that took care of presenting data, reading responses, and handling mouse input and menu display.

Since most interactive programs need to be able to do certain things, Symbolics decided to provide a *substrate* layer which provided easy ways to do them. This substrate includes:

- A generic read/interpret/redisplay command loop which works for any interactive system.

- A mechanism for displaying many different types of data on the screen, and having the system remember the object and its underlying type.

- A means of dividing up screen "real estate" into a

programmer-controlled *framework*, allowing the programmer to specify relative or absolute sizes of the divisions,
the types of display in each, and so forth.

- An interactive "layout designer" program which aids in
  the construction of program frameworks.

Most of this is well-documented in volume 7A of the Symbolics
documentation set. This chapter is an overview and a few suggestions for some things which work better than others.

6.1 Program Frameworks: an Overview

A *Program Framework* is the user interface nexus for
medium- and large-scale interactive systems. The idea comes
from the following observation:

> Most interactive programs consist of an "infinite"
> loop, consisting of command reading, command execu-
> tion, and display update. This includes such primi-
> tive interactors as the Lisp Listener, and as sophis-
> ticated ones as the text editing and mail reading
> programs.

The philosophical model is similar to that presented in *Presentation Based User Interfaces*, a PhD thesis written by Gene Cic-
carelli at the MIT AI Lab. A "database" underlies the system,
the display presents its current state, and commands are used
to modify the database or the user's view of it.

A program framework is a way to organize the following four
related items:

1. A window-layout declaration,
2. A set of window display definitions,

3. A command loop, and
4. A set of commands.

A program framework is defined using the macro **dw:define-program-framework**. It contains the name of the program and a large number of options. Among other things, **dw:define-program-framework** creates a flavor whose name is the same as the name of the program. Here are some of the interesting options:

- **:panes** – lists the window panes which divide up the real estate of the total window. Each pane has a name, a type, and some per-type redisplay information.

- **:configurations** – describes how the panes are actually laid out on the screen.

- **:select-key** – defines the character which can be used after Select to obtain this program window.

- **:system-menu** – declares that this program is to appear in the third column of the system menu.

- **:state-variables** – a list of instance variables of the flavor defined by **dw:define-program-framework**. All these instance variable definitions must contain initialization values. All commands will be methods of the flavor, so these state variables will be accessible from command bodies.

- **:command-definer** – declares that this program has a command definer. The option may be the name of the command definer. It may also be **t**, which means to define a command defining macro named **define-*program-name*-command**. If this option is not specified, or is **nil**, no command definer is provided.

There are several types of panes. Here are some of the common ones:

- **:interactor** – A pane for interactive input/output. This will hold your command history. [A **:listener** is the same, except that it is taller and standard I/O variables like ***query-io*** are bound differently]

- **:display** – A pane for output display. This is where your database display might be done. You might have several **:display** panes in your window.

- **:title** – A constant display telling the user what the window is for.

- **:command-menu** – A menu of commands, mouse-sensitive.

Pane options are used to define the size and shape of the pane and its displayed contents. Here are some of the latter:

- **:redisplay-function** – a function which is called to produce the output. It is passed two arguments, namely the program instance and the window pane itself. A good way to write one of these is to make it be a method on the program flavor, which allows you to have access to its state variables.

- **:redisplay-after-commands** – if non-**nil**, the redisplay function is called after every command the user types. If **nil**, it's only called when the window is first created, and when the entire frame is refreshed.

- **:incremental-redisplay** – if **nil**, the window's output history is cleared before your redisplay function is called; this has the effect of erasing all the contents of the window. If this is **t**, your redisplay function is called as a redisplayer, and is expected to use the standard incremental redisplay technology. If it's the keyword **:own-redisplayer**, your function is supposed to do its own incremental redisplay, using some other technique to remember what needs to be updated (the practical effect of this value is that the **:clear-history** message is *not* sent to the window before your function is called).

- **:menu-level** – is for command menus. It specifies which set of commands goes in this menu. It's for programs that want to segregate their commands into more than one menu, for example, for commands that affect groups of objects and commands that affect single objects in your database. The command definitions also have **:menu-level** declarations in them.

- **:name** – Sets the name you can type to invoke this command. One use of this option is **:name nil**, which says that you can't type this command. If you do this, you should supply some other way to invoke the command, such as a menu accelerator.

Sample program framework definitions are included in each of the example programs in this book. Symbolics also ships two example program files with the system source:

- SYS:EXAMPLES;DEFINE-PROGRAM-FRAMEWORK.LISP
- SYS:EXAMPLES;UI-APPLICATION-EXAMPLE.LISP

6.2 Defining Commands

Program framework commands are defined using a macro created by the **dw:define-program-framework** macro. If you specify **t** for your program's command definer, the name of the command definer will be **define-*program-name*-command**; otherwise, it will have whatever name you've specified in that position.

Regardless of its name, its syntax is very much like the CP's command definition macro, **cp:define-command**. The first operand is a list which has the name of the command and various keyword options; the next is a list of arguments to the command, whose presentation types you declare in the argument list. These are followed by the body of the function which defines the command. Command bodies are always methods of the flavor defined for your program, which means that they can access the state variables as instance variables.

In addition to the **cp:define-command** keywords allowed in the first subform, program command definitions are permitted several extra keywords.

:menu-accelerator

>The string which appears in the menu display for this command.

:menu-level   The "level" of the menu in which this command is to appear.

:keyboard-accelerator

>The one-character abbreviation for this command. This only works well if you specify that the **:command-table** for your program framework has a keyboard accelerator *and* you read your commands without echoing them by default. See the chapters 9 and 11 for how this is done.

6.3 The Redisplay

Each of your :**display** panes may have a redisplay function. It is responsible for the entire display content of the pane. By default, all panes have a display history which remembers everything you've printed on it. Also by default, however, every pane's output history is cleared just before your redisplay function is called, so your function must put everything your user cares about into the window each time.

The way to avoid this (and, in the process, avoid driving your users crazy) is with incremental redisplay. There are two different ways to do incremental redisplay.

1. You can use the system-supplied one (this is what the calculator example does in chapter 9). Say :**incremental-redisplay t** as part of the pane description. All your output should use the redisplayable-output facilities described in volume 7A (See the section "Redisplay Facilities" in *Programming the User Interface, Volume a.*). For example, try using **dw:independently-redisplayable-format**.

2. You can "roll your own," and remember what's supposed to be redisplayed (this is what the card-game example does in chapter 11). Say :**incremental-redisplay :own-redisplayer** in the pane description. The primary effect of :**own-redisplayer** is to suppress the clearing of the pane's output history. [You will have to do this if you are trying to use overlapping output, because the system-supplied one doesn't perform overlapping output in a useful way.]

Another less-obvious way to do redisplay is to have your commands to it explictly. To get your hands on the pane involved, you need to use **dw:get-program-pane**, which I recommend you look up.

6.4 Presentation Types

Lisp objects have an inherent type, which the Lisp Machine architecture keeps track of. Additionally, the Common Lisp type system allows the programmer to specify complicated types, as well as simple ones. For example, a Common Lisp program may specify that its first argument must be "an integer between 0 and 100," or must be "*either* a positive rational number *or* nil." The way this is done is by specifying *data-type* arguments as part of the type expression.

However, an object's extrinsic meaning is whatever the programmer assigns to it. For example, the system keeps the time of day in the form of an integer which records the number of seconds elapsed since January 1, 1900 at midnight GMT. If you just print out that integer as an integer, you lose the meaning that number had to the programmer, and it becomes just an integer. On the other hand, if you print it as a time, then it means a completely different thing to the reader. Suddenly, that integer has more *semantic information* attached to it.

This is what *presentation types* are for. They describe the translation between internal representation and printed representation at a higher level than the Lisp type system can (they are an extension to that system). Just as with Common Lisp types, they have data-type arguments which can augment or diminish the class of objects they describe. In addition, there is another dimension along which data displays can vary: they may be *presented* differently. For example, the number represented by "100" in base 10 is the same as the number represented by "144" in base 8, "64" in base 16, and "121" in base 9. So, in addition to data-type arguments, presentation types have *presentation arguments*.

The two functions which use presentation types are **accept** and **present**. As with **read** and **print**, these functions convert between what the user types or reads and what the system stores internally. They each take a presentation type as an argument; **present** also, obviously, takes an object to present. Some sample uses:

```
(present 100 'integer) -> 100
(present 100 '((integer) :base 8)) -> 144
(accept 'integer) 100 -> 100.
(accept '((integer) :base 8)) 100 -> 64.
(accept '((integer 0 10) :base 8) 15 -> error
                                  10 -> 8.
```

As you can see, a presentation type is either:

- A symbol

- A list. The first element of the list is a list which is a Common Lisp type, with optional data-type arguments. The rest of the list is a (potentially empty) list of presentation arguments.

The system supplies a number of presentation types by default. Many of them have very specialized applications, but a number of them are for general use. Some simple ones:

boolean   Accepts "Yes" or "No", returns **t** or **nil**

keyword   Accepts any string, returns a keyword symbol by that name.

character   Accepts the first character you type, and returns that character.

pathname   Accepts a typed-in file pathname.

time:universal-time
                      Accepts a time of day, returns an integer.

time:time-interval
                      Accepts a time interval (*e.g.*, "3 weeks"),
                      returns the number of seconds in that inter-
                      val.

net:host   Accepts the name of a computer in the net-
                      work database.

sys:printer   Accepts the name of a printer in the network
                      database.

sys:form   Any lisp input form.

Here are some which take required data-type arguments and
return one or more of them:

member   returns one of the following objects.

subset   returns zero or more of the following objects.

alist-member   Accepts a string, returns the associated ob-
                      ject.

Here are some whose data-type arguments are more presentation types:

token-or-type   Accepts one particular token, or any element
                      of another presentation type.

null-or-type   Special case of the above, takes "None" or
                      another type.

sequence-enumerated
                      Accepts the following presentation types, in
                      order, and returns a list of values.

or   Accepts any one of the following presentation types.

You can write your own presentation types. See the section "Presentation-Type Definition Facilities" in *Programming the User Interface, Volume a.* Also, I've defined a number of presentation types in the sample programs in chapters 7, 9 and 11

6.5 Mouse Sensitivity

We say that a displayed item on the screen is *mouse sensitive* when pointing at it with the mouse has some defined meaning: it will cause an action, or indicate a choice. Every displayed output on a window is potentially mouse-sensitive.

In previous releases of Lisp Machine software, making mouse-sensitive displays on the screen was very tedious, and fraught with dangers. Writing software which did mouse-sensitive output required learning a great deal about the innards of the window system. In general, each program which wanted to do such output required one or more new flavors of window, each with some methods which ran in the mouse process. If there were any bugs in your software (and there *always* were), you spent a lot of debugging time in the cold load stream, or, worse yet, crashing your machine.

No longer. Most programmers need only learn to make one kind of window, namely a *Dynamic* window.[1] Dynamic Windows remember all of the following for each item of output (in addition to its location on the window):

---

[1] The flavor of window is **dw:dynamic-window** or a flavor built on it, as are, for example, the panes in program frameworks.

1. What object the display represents,
2. What kind of display it is (its *presentation type*), and
3. Other display-related options (e.g., what base it was displayed in, whether a list was displayed as code, a property list or data, etc.).

Note that *which* presentations are mouse-sensitive at any given time depends on two things:

1. The *input context*. If you are using **accept**, for example, only those presentations which are of the type you are accepting (or can be translated into that type: See the section "Mouse Gesture Translations," page 144.) are mouse-sensitive.

2. The *shift keys* which are pressed at the time. "Shift keys" includes not only the Shift button, but also Control, Meta, Super, Hyper, or any combination of them.

One other thing to keep in mind: Dynamic windows remember *all* the output ever done on a window (until you clear its *output history*). This includes the output which has scrolled off the top of the window. Dynamic windows can be scrolled forward and back (and left and right, if any of your output moves over the right-hand edge of the window) to show output which has scrolled off the screen. Any output which was presented with an appropriate type, even if scrolled back onto the screen, will be mouse sensitive at the appropriate times.

6.5.1 Mouse Sensitivity – the Easy Part

All output on Dynamic windows is mouse-sensitive. Depending on how much control you want over the presentation type of the display, you have three options for facilities with which to present the output:

1. Any printing operation, e.g., **print** or **format**, presents its output as simple presentation types. When you use **print**, for example, on a lisp object, it is presented using its data type (as defined by **type-of**) as its presentation type.

2. The **present** function permits you to specify the exact presentation type you wish to use. You might use this, for example, if you have an object which *is* an integer, but *represents* a universal time. In general, **present** is for objects whose data type doesn't tell the whole story.

3. The macro **dw:with-output-as-presentation** permits you to specify the data type *and* the manner in which the data is presented on the screen. All output drawn on the screen inside one of these forms becomes part of the presentation. You can use this to make random[2] graphics be part of your presentation.

Once you have presented your object, you probably want to read it somehow. Mouse sensitivity is controlled entirely by what your program is attempting to read at the time, i.e., its *input context*.

Just as there are three ways of presenting output, depending on how much control you want to have over its appearance and underlying type, there are three corresponding input mechanisms:

1. Any input operation, e.g., **read**, accepts its input in the form of mouse clicks on sensitive items. The command processor also accepts both whole commands and single arguments to those commands; thus previously-typed commands and other output are often sensitive while the command procesor is waiting for you to type a command.

---

[2]See the *Fun and Games* section at the end of this chapter.

2. The function **accept** is used to control this behaviour
   more precisely. You can specify not only the presentation
   type but its data arguments and presentation arguments.
   accept is especially powerful inside a
   dw:accepting-values form, where several calls to **accept**
   become a single *menu* of choices.

3. The macro **dw:with-presentation-input-context** allows the
   user to control exactly what input is sensitive while performing any arbitrary input operations. This is the most
   flexible input control you want: inside its body, you
   specify what the keyboard-reading operation is, and also
   what to do when a mouse operation is performed.

6.5.2 Mouse Gesture Translations

As I hinted earlier, a presentation is mouse-sensitive in a number of contexts. When you're accepting input of its presentation type, obviously it's mouse sensitive. Similarly, when
you're accepting input of a type which is a superset of the type
used to present the object, *that* presentation is also mouse-
sensitive. For example, presentations of type **integer** are sensitive when you're accepting **numbers**.

Another case is where the type you're accepting is a subset of
the type used to present the object, *and* the object happens to
fall into that subset. For example, if you are accepting integers between 0 and 9, the output displayed by (present 3
'number) is mouse sensitive.

The final context in which a display is mouse-sensitive is when
you (or the system) has defined a *translation* between the type
of the display and the presentation type given to **accept**. For
example, there is a system-defined translation which converts
pathname presentations to the *Show File* command. If the
machine is waiting for you to type a command, you can click on

a pathname display, and it will be just as if you had typed Show File *pathname.*

There are many such translations already defined in the system. They are far too numerous to list here. You can discover many of them by clicking right on presentations and looking at the menu provided. Also, clicking right on presentations while holding down the Super key will give you a great deal more to explore via menus.

Presentation translators specify a "from" data type and a "to" data type, and a function used to convert between them. In addition, the programmer may supply a number of other options:

:tester   A function which may be used to determine whether the object is interesting to the user in this context.

:gesture   Which mouse gesture invokes this translation.

:priority   Whether this translator is more important than other potentially selectable ones.

:suppress-highlighting
              Whether drawing boxes around the sensitive objects ought to be suppressed.

There are other options. For complete documentation: See the macro **define-presentation-translator** in *Programming the User Interface, Volume a.*

One variant of **define-presentation-translator** which is worth mentioning is **define-presentation-to-command-translator**. **define-presentation-to-command-translator** is used in the same manner as **define-presentation-translator**, except that its result is always a command. For example, the translator described above which converts pathnames to the *Show File* command is defined as follows:

```
(define-presentation-to-command-translator
  si:com-show-file              ; The name of the translator
  (fs:pathname)                 ; What it translates from
  (file)                        ; The argument to the body
  '(si:com-show-file (,file)))  ; The body of the translator
```

Here is another example. What this one does is to offer the command Load File when the file is a compiled object file. While it has a higher priority than the previous one, its tester keeps it from being used on any files which you might want to print out.

```
(define-presentation-to-command-translator
  si:com-load-binary-file
  (fs:pathname
    :gesture :select
    ;; boost it over Show File when tester succeeds
    :priority 0.5
    :tester ((path)
             (eq (send path :canonical-type) ':bin)))
  (path)
  (cp:build-command 'si:com-load-file (ncons path)))
```

[There is also another whole class of mouse-sensitivity definitions, called *presentation actions*. These are defined using **define-presentation-action**. They define side effects the user might like to cause while the system is waiting for input. An example of the kind of thing a presentation action is used for is clicking c-m-Middle on structure and instance slots to replace their contents.]

6.6 Fun and Games

From *The Hacker's Dictionary*, Guy L. Steele, Jr., *et al*:

RANDOM *adj.*

1. Unpredictable (closest to mathematical definition); weird. "The system's been behaving pretty randomly."
2. Assorted; undistinguished. "Who was at the conference?" "Just a bunch of random business types."
3. Frivolous; unproductive; undirected (pejorative). "He's just a random loser."
4. Incoherent or inelegant; not well organized. "The program has a random set of misfeatures." "That's a random name for that function." "Well, all the names were chosen pretty randomly."
5. Gratuitously wrong, i.e., poorly done and for no good apparent reason. For example, a program that handles file name defaulting in a particularly useless way, or a routine that could easily have been coded using only three ac's, but randomly uses seven for assorted non-overlapping purposes, so that no one else can invoke it without first saving four extra ac's.
6. In no particular order, though deterministic. "The I/O channels are in a pool, and when a file is opened one is chosen randomly."
7. *noun.* A random hacker; used particularly of high school students who soak up computer time and generally get in the way.
8. *(occasional MIT usage)* One who lives at Random Hall.

J. RANDOM is often prefixed to a noun to make a "name" out of it (by comparison to common names such as "J. Fred Muggs"). The most common uses are "J. Random Loser" and "J. Random Nurd" ("Should J. Random Loser be allowed to gun down other people?"), but it can be used just as an elaborate version of RANDOM in any sense. [See also the note at the end of the entry for **HACK**.]

RANDOMNESS *noun.* An unexplainable misfeature; gratuitous inelegance. Also, a hack or crock which depends on a complex combination of coincidences (or rather, the combination upon which the crock depends). "This hack can output characters 40-57 by putting the character in the accumulator field of an XCT and then extracting 6 bits -- the low two bits of the XCT opcode are the right thing." "What randomness!"

7. The Graph Example

This chapter, rather than present some abstracted features of the lisp language or the lisp machine operating environment, will cover a programming example which puts to use many of the features we have previously discussed. The piece of code in question allows one to display and manipulate simple undirected graphs, that is, sets of *nodes* connected by *arcs*.

If your site has loaded the tape which accompanies this book, you can load the code by using the CP command Load System grapher. Once the code has been read, start the program by typing Select Circle.[1] The program window will look something like figure 4.

This chapter contains a listing of the program. The first three sections will point out and briefly discuss the interesting features of the code. All of the program's files live under the logical directory LISP-LORE:EXAMPLES;GRAPHER;. The files are:

---

[1]The "Square," "Circle" and "Triangle" keys on the top of the keyboard are reserved for any application the user wants. Symbolics will never assign meanings to them.

- NODES-AND-ARCS — the data abstraction
- PRESENTATION-TYPES — the translations between data objects and display presentations
- DISPLAY-FRAME — the program framework and commands

7.1 The Nodes and Arcs

There are two data types defined in this program. The first is called **node,** and the second is **arc.** A list of **nodes** is kept in a global (i.e., special) variable. The list of **arcs** is *implicit* – it is derived from the list of **nodes.** Some notes:

- The two **defvars:** These declarations are for global variables that will be needed at various places throughout the code. A **defvar** must precede the first use of the variable so that the compiler knows the symbol refers to a *special variable.* Another good reason for putting them at the beginning is so anyone reading the code can quickly find out what hooks are available for getting their hands on the program's internal data representation.

- The **node** flavor: Four of the instance variables are initable, meaning that you can specify their values when you create a node. The **radius** and **unique** instance variables are *internal* state not directly visible to the user. The **label** and **shape** instance variables may be set (and read) from outside the instance; in addition, the list of **arcs,** and the node's position and radius may be read.

- The **make-instance** and **sys:print-self** methods: The last thing **make-instance** always does is to call the flavor's **make-instance** method, if it has one. Here, you can specify operations to be performed on every instance of your flavor, upon being initialized. I use this one to add the node to the list of all nodes.

Figure 4.   Grapher program display window

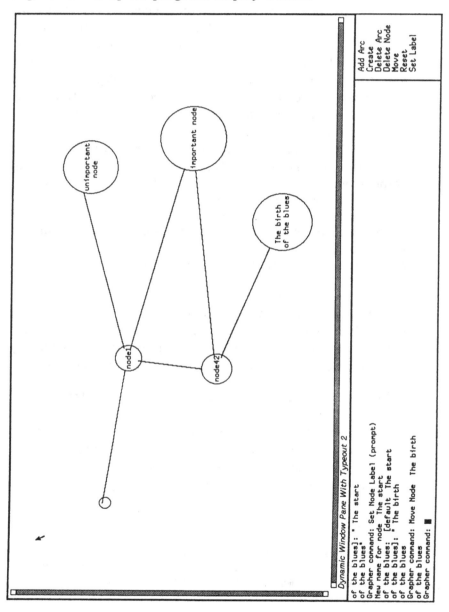

The functions **print, princ, format** and so forth all use the **sys:print-self** method of instances. The first argument (after the instance itself) is the stream on which to print the result, the second (not used here) is ***print-level*** (as modified for the current printing depth), and the third is whether to print *readably*.[2] This last means that **read** should be able to reconstruct the object from the characters you print. Since this is impossible for most flavor objects, we don't even try to do this. **sys:printing-random-object** is a macro which, among other things, prints the "#<" and ">" around most instances when they are printed out. The reader will signal an error when it tries to read "#<."

- **ignore** as an argument: Use of **ignore** in a lambda-list for an argument which isn't going to be used saves you from getting a compiler warning about an unused variable.

- **map-over-nodes**: This macro is used to iterate over all the nodes. While it is not hard to write the loop in the few places where it is necessary, I did it for symmetry with the macro **map-over-arcs**, which is discussed below.

- **:after** methods for **setf**: The methods constructed for setfing the instance variables declared with **:writable-instance-variables** are just like any other methods, and can have daemons and/or whoppers. For more sophisticated applications, you can have special method combinations, like **:case** (which might be used, for

---

[2] You will sometimes see this called **slashify-p**, which means that you are supposed to print "\" characters before all the special characters which need to be quoted. I prefer **readably**, which indicates a little better what is going on.

example, if there were only a small number of things you wanted to set the instance variable to, and you wanted to do something different for each of the possible values).

In this case, I used the **:after** method to write down the fact that the previously calculated radius is likely to be wrong, and needs to be recalculated.

- The **move** method: Remember that we didn't declare the **x-pos** and **y-pos** instance variables to be writable? This is a small lesson in modularity. We want to say that the only way to specify the position of a node is to "move" it to a new position. If we wanted to attach other actions to moving the node, we could create daemons or whoppers to the **move** method; to do this for, say, **(setf y-pos)** as well would be a potential modularity violation, in that we would have to duplicate the code in more than one place.

- The **present-self** method: This is the output side of the mouse-sensitive display of nodes. Note that there are two versions of this in the text; the commented-out one is the one I thought would be correct, but doesn't produce the correct result. The other one uses a clumsier mechanism for producing the output, but makes the dynamic window database be correct.

This method uses **dw:with-output-as-presentation** to make the output be a *presentation*. All of the output within its contour is a single presentation "box", whose presentation type is **node** and which refers back to the **node** object itself.

Inside that macro, we say we want to surround the output with whatever shape is appropriate using **surrounding-output-with-border**. Inside *that*, we finally produce the output, which is the label, if specified, or a blank string.

- **map-over-arcs**: I wrote this macro to have a handy way to iterate over all the arcs. For each node, it runs through all its arcs. If it has already seen that arc it goes on. If it hasn't, it executes the body forms with the specified variable bound to the arc, and marks the arc as visited. Note the use of **make-symbol** to ensure the mark variable and node variables don't shadow any variable bindings.

- The rest of the methods for **arc**: These are all parallel to those for **node**. Note that presenting **arcs** doesn't actually **present** them, because the Genera 7.0 display substrate isn't powerful enough to allow mouse-sensitive lines (they become mouse-sensitive rectangles big enough to surround the entire line, but also a lot more screen area). This is supposed to be fixed for a later release.

7.2 The Presentation Types

Since we only **present** the nodes, there is only one presentation type in the file. Some notes:

- It says **:no-deftype t** because there is already a "type" by the name **node**. What is it? The flavor, of course. The "type" namespace contains flavors (from **defflavor**), structures (from **defstruct**) and Common Lisp types (from **deftype**), as well as presentation types.

- The parser appears to be an infinite loop. Fortunately, it's called in an environment where the system is looking for mouse clicks. I call this parser the *null parser*; since you have to supply a parser whenever you're going to actually accept an object of the given type, this parser is the one I use when you can't type one in, and must therefore click on one.

7.3 The Display

The program framework for this program is called **grapher**. Some notes:

- The **:select-key**, **:command-definer** and **:command-table** keywords are well-documented, and not discussed here.

- The **:state-variables**: none are supplied. To do this program correctly, the data kept in the global variable ***all-the-nodes*** should really be kept in a state variable for the program instead. What this means is that if a user creates more than one grapher frame at a time (with, for example, Select Control-Circle), they will interact incorrectly. I have left this cleanup as an exercise for the reader. See the section "Problem Set," page 169.

- The **graph** pane is the interesting one. It is a **:display** pane, which means that the only display output it gets is what your program does to it. Here's what the rest of the keywords mean:

  ° **:redisplay-after-commands** means that every time the user types a command (or performs one with the mouse), the redisplay function will be called.

  ° **:redisplay-function** is how you specify what function actually does the redisplay. The function takes two arguments, the program instance and the pane on which it is to be called. One way to write this function is as a method for the program flavor; this allows you to use the state variables as instance variables.

  ° **:incremental-redisplay** is how you would specify

that you were only going to update part of the window instead of the whole thing. This program
clears the screen and displays the entire graph over
again from scratch after every command.

° **:margin-components** allows you to put things in the
margins of the window. The margin components actually in use are listed at the beginning of the file.
You get "ragged" borders whenever you've scrolled
some of the display off the screen, scroll bars in the
left margin and the bottom margin (the mouse
documentation talks about scrolling parts of the
"display"), and a little white space just inside the
scroll bars.

° **:typeout-window** means that typing Suspend or
entering the debugger will happen on a typeout window in this pane.

• The **:configurations** keyword supplies the layout and
proportions of the panes declared in the **:panes** list. Both
the **:panes** and **:configurations** lists were written by
"Frame-Up", the program you get to with m-X Create
Program Framework in Zmacs, or <Select>Q. I then edited
the panes by hand to add the margin components and
change the order; I also converted it to lower-case, because I prefer to read and write my code in lower-case.

The method **display-graph** presents all the nodes on the display pane, followed by all the arcs. It is done in this order because displaying an arc requires calculating the radii of the
nodes involved, and presenting a node always calculates its
radius.

7.4 The Commands

The commands are all pretty straightforward. They are defined using **define-grapher-command**, the macro written in response to **:command-definer t** in the program framework definition.

- **com-create-node** creates nodes. It can be typed as Create Node <integer> <integer>. You can also click on "Create" in the menu.

- **com-set-node-label** takes a **node** as an argument. Remember that you can't type them in; this means that you have to click on a node in order to pass it as an argument to this command.

- **com-move-node** tracks the mouse until you click in the place where you want to move the node.

- **com-add-arc** creates an arc. Again, you have to click on both nodes involved.

- **com-delete-node** and **com-delete-arc** work in the obvious way. In **com-delete-node**, note the use of "˜⊂" and "˜⊃" in **format** control strings to delimit character style changes.

- **com-reset-database** clears the database of nodes.

7.5 The Mouse Gesture Translators

There are several translators which convert mouse gestures into commands. Some of them are not the most elegant ways that you might imagine were correct, but they are the mechanisms I have found through trial and error. The bindings between gesture names and the mouse characters which invoke them is at the end of the file.

- **com-create-node-ex-nihilo** translates a "create node" gesture on an empty place in the frame into a node-creating command.

- **com-set-node-label** translates a "set node label" gesture on a node into a call to a special internal command which prompts for a label, and sets it.

- **com-move-node-from-here-to-there** translates a "move node" gesture on a node into a "Move Node" command.

- **com-add-arc-from-here-to-there** and **com-delete-arc-from-here-to-there** translate "add arc" and "delete arc" gestures, respectively, on a node into special internal commands which accept a click on a second node and create or delete an arc from one to the other.

- **com-delete-node** translates a "delete node" gesture into the command which deletes a node.

- The **define-grapher-mouse-gestures** form at the end actually connects each gesture to its respective mouse click. It uses **setf** on **dw:mouse-char-for-gesture** to do this.

7.6 The Program

lisp-lore:examples;grapher;grapher-system.lisp

```
;;; -*- Mode: LISP; Syntax: Common-lisp; Base: 10 -*-

(defpackage grapher
  (:use scl)
  (:colon-mode :external)
  (:export *all-the-nodes*))

(defsystem grapher
    (:patchable t
     :default-pathname "lisp-lore:examples;grapher;"
     :maintaining-sites :ssf
     :pretty-name "Graph demo program")
  (:serial "nodes-and-arcs" "presentation-types" "display-frame"))
```

lisp-lore:examples:grapher:nodes-and-arcs.lisp

```
;;; -*- Mode: LISP; Syntax: Common-lisp; Package: GRAPHER; Base: 10 -*-
;;; Node definitions

(defvar *all-the-nodes* nil
  "a list of instances of flavor node.
   only active nodes appear here")

(defvar *next-node-index* 0
  "Unique index assigned to each node.")


;;; The flavor of each node in the graph.
(defflavor node
        ((arcs nil)                              ; list of arcs attached to this node
         xpos                                    ; coordinates of the center of this node
         ypos
         (radius :needs-calculation)             ; radius of node
         (label nil)                             ; a string
         (shape :circle)                         ; what shape border is displayed
         (unique (incf *next-node-index*)))      ; For print-self
        ()                                       ; No other flavors
     ;; You can set location, label and shape instance variables when you create
     ;; an instance of node. After creating, you can change the label and shape.
     ;; You can always read the list of arcs, position and radius, in addition to
     ;; label and shape. The position must be specified when creating a node.
     (:initable-instance-variables xpos ypos label shape)
     (:writable-instance-variables label shape)
     (:readable-instance-variables arcs xpos ypos radius)
     (:required-init-keywords :xpos :ypos))


;;; Called when new instances are created
(defmethod (make-instance node) (&rest ignore)
  (push self *all-the-nodes*))                   ; Record each active instance


;;; Print-self routine to make nodes print prettily.
(defmethod (sys:print-self node) (stream ignore readably)
  (let ((name (or label (format nil "Unnamed node ~D" unique))))
    (if readably                                 ; Print so read can "work".
        (sys:printing-random-object (self stream
                                          :typep
                                          :no-pointer)
          (write-string name stream))
        (write-string name stream))))
```

```
;;; For performing an operation to all nodes.
(defmacro map-over-nodes ((node-var) &body body)
  `(loop for ,node-var in *all-the-nodes*
         do (progn ,@body)))

;;; Need to recalculate radius
(defmethod ((setf node-label) node :after) (ignore)
  (setf radius :needs-calculation))

;;; Move a node from here to there
(defmethod (move node) (new-xpos new-ypos)
  (setf xpos new-xpos ypos new-ypos))

;;; Self-explanatory
(defmethod (add-arc node) (arc)
  (push arc arcs))

(defmethod (remove-arc node) (arc)
  (setq arcs (delete arc arcs)))
```

```
;;; Remove a node from the database
(defmethod (delete-self node) ()
  ;; (borrowed from tv:sheet's :kill method)
  ;; Do it this way to prevent CDR'ing down list structure being modified
  (loop until (null arcs)
        do (delete-self (car arcs)))
  (setq *all-the-nodes* (delete self *all-the-nodes*)))

;;; Display a node on the given window
(defmethod (present-self node) (window)

  (calculate-radius self window)              ; Only required because we want to center
                                              ; label where mouse click happened
  ;; This is written this way because of a bug in 7.0 -- correct code below
  (send window :set-cursorpos (- xpos radius) (- ypos radius))
  (dw:with-output-as-presentation (:stream window :object self
                                   :type 'node :single-box t)
    (surrounding-output-with-border (window :shape shape)
      (send window :string-out (or label " ")))))

#| Correct (i.e., Ideologically pure) version of the above:

;;; Display a node on the given window
(defmethod (present-self node) (window)
  (calculate-radius self window)                ; Only required because we want to center
                                                ; label where mouse click happened
  (dw:with-output-as-presentation (:stream window :object self
                                   :type 'node :single-box t)
    (surrounding-output-with-border (window :shape shape)
      (graphics:draw-string (or label " ") (- xpos radius) ypos :stream window))))

|#

;;; Calculate the radius of a node for a given window.
;;; We only care about this because we want to center the node at (xpos, ypos)
;;; instead of having its upper left be there.
(defmethod (calculate-radius node) (window)
  (when (and (stringp label) (zerop (string-length label)))
    (setq label nil))
  (when (eql radius :needs-calculation)
    (setq radius (ceiling
                   (multiple-value-bind (width height)
                       (dw:continuation-output-size
                         (lambda (stream)
                           (surrounding-output-with-border
                             (stream :shape shape)
                             (send stream :string-out (or label " "))))
                         window)
                     (max width height))
                   2)))
  radius)
```

```
;;; Arc definitions
;;; The flavor of each arc in the graph
(defflavor arc
        ((mark nil)    ; arbitrary object for misc tagging purposes
         node1         ; the two "node" objects this arc connects
         node2)
        ()
  (:writable-instance-variables mark)
  (:initable-instance-variables node1 node2)
  (:readable-instance-variables node1 node2))

;;; Hook this arc up to its nodes
(defmethod (make-instance arc) (&rest ignore)
  (add-arc node1 self)
  (add-arc node2 self))

;;; Arcs always print in #<...> format.
(defmethod (sys:print-self arc) (stream &rest ignore)
  (sys:printing-random-object (self stream :typep)
    (format stream "~A ←→ ~A" node1 node2)))

;;; Macro to perform an operation on all arcs.
;;; Note that this cannot be nested inside itself.  Thus, it can't be used
;;; to create relations whose domain and range are the set of arcs, for example.
(defmacro map-over-arcs ((arc-var) &body body)
  (let ((mark-var (make-symbol "MARK"))
        (node-var (make-symbol "NODE")))
    `(loop with ,mark-var = (ncons nil)          ; Create mark
           for ,node-var in *all-the-nodes*
           do (loop for ,arc-var in (node-arcs ,node-var)
                    unless (eq (arc-mark ,arc-var) ,mark-var)
                    do (progn ,@body
                              (setf (arc-mark ,arc-var)
                                    ,mark-var))))))

;;; Self-explanatory
(defmethod (delete-self arc) ()
  (remove-arc node1 self)
  (remove-arc node2 self))

;;; Currently, no good way to present arcs, just draw them instead.  We can
;;; count on the radius being OK because we always draw the nodes before the
;;; arcs; otherwise we'd have to call calculate-radius on them.
(defmethod (present-self arc) (window)
  (multiple-value-bind (x1 y1 x2 y2)
      (find-edges-of-nodes (node-radius node1)
                           (node-xpos node1)
                           (node-ypos node1)
                           (node-radius node2)
                           (node-xpos node2)
                           (node-ypos node2))
    (graphics:draw-line x1 y1 x2 y2 :stream window)))

;;; Subroutine for the above.
(defun find-edges-of-nodes (r1 xpos1 ypos1 r2 xpos2 ypos2)
  (let* ((dx (- xpos2 xpos1))
         (dy (- ypos2 ypos1))
         (length (isqrt (+ (* dx dx) (* dy dy)))))
    (values (+ xpos1 (ceiling (* dx r1) length))
            (+ ypos1 (ceiling (* dy r1) length))
            (- xpos2 (floor (* dx r2) length))
            (- ypos2 (floor (* dy r2) length)))))

(compile-flavor-methods node arc)
```

lisp-lore:examples:grapher:presentation-types.lisp

```
;;; -*- Mode: LISP; Syntax: Common-Lisp; Base: 10; Package: GRAPHER -*-

;;; This is the presentation type for nodes
(define-presentation-type node
        ()
    :no-deftype t                      ; node is also a flavor
    ;; Standard parser for "you can't read one of these."
    :parser ((stream) (loop do (dw:read-char-for-accept stream))))

#| We're not presenting any of these at the moment.

;;; This is the presentation type for arcs
(define-presentation-type arc
        ()
    :no-deftype t                      ; arc is also a flavor
    ;; Standard parser for "you can't read one of these."
    :parser ((stream) (loop do (dw:read-char-for-accept stream))))

|#
```

lisp-lore:examples:grapher:display-frame.lisp

```lisp
;;; -*- Mode: LISP; Syntax: Common-Lisp; Base: 10; Package: GRAPHER -*-

(defparameter
  *grapher-display-margin-components*
  'dw:((margin-ragged-borders )
       (margin-scroll-bar :history-noun "display")
       (margin-scroll-bar :history-noun "display"
                          :margin :bottom)
       (margin-white-borders :thickness 2)))

(dw:define-program-framework grapher
  :select-key
  #\Circle
  :command-definer
  t
  :command-table
  (:inherit-from '("colon full command" "standard arguments" "standard scrolling"))
  :state-variables
  nil
  :panes
  ((graph :display :redisplay-after-commands t
          :redisplay-function 'display-graph
          :incremental-redisplay nil
          :margin-components *grapher-display-margin-components*
          :typeout-window t)
   (commands :interactor)
   (menu :command-menu :menu-level :top-level :columns 1))
  :configurations
  '((dw::main
      (:layout (dw::main :column graph row-1)
       (row-1 :row commands menu))
      (:sizes (dw::main (graph 0.8) :then (row-1 :even))
       (row-1 (menu :ask-window self :size-for-pane menu)
        :then (commands :even))))))

(defmethod (display-graph grapher) (display-pane)
  (map-over-nodes (node) (present-self node display-pane))
  (map-over-arcs (arc) (present-self arc display-pane)))

(define-grapher-command (com-create-node :menu-accelerator "Create")
    ((x 'integer)
     (y 'integer))
  (make-instance 'node :xpos x :ypos y))

(define-presentation-to-command-translator com-create-node-ex-nihilo
                               (dw:no-type :blank-area t
                                           :suppress-highlighting nil
                                           :documentation "Create node here"
                                           :gesture create-node-ex-nihilo)
  (ignore &key x y) (cp:build-command 'com-create-node x y))

(define-grapher-command (com-set-node-label :menu-accelerator "Set Label")
    ((node 'node :prompt "a node")
     (label '((null-or-type string))
            :default (node-label node)
            :prompt "its label" :confirm t))
  (setf (node-label node) label))
```

```
(define-grapher-command (com-set-node-label-\(prompt\) :name nil)
    ((node 'node))
    (let* ((label (accept 'string
                          :prompt (format nil "New name for node ~A: " node)
                          :default (node-label node))))
       (setf (node-label node) label)))

(define-presentation-to-command-translator com-set-node-label
                                            (node :gesture set-node-label)
  (node)
  (cp:build-command 'com-set-node-label-\(prompt\) node))

(defmacro with-mouse-tracking-node ((node pane) &body body)
  ;; Done right, we would care about the node's radius
  (let ((old-mouse-char (make-symbol "OLD-MOUSE-CHAR")))
    `(let ((,old-mouse-char (send ,pane :mouse-blinker-character)))
       (unwind-protect
           (progn (send ,pane :set-mouse-cursorpos
                              (node-xpos ,node) (node-ypos ,node))
                  (send ,pane :set-mouse-blinker-character
                              #\mouse:fat-circle)
                  (send ,pane :mouse-standard-blinker)
                  ,@body)
         (send ,pane :set-mouse-blinker-character ,old-mouse-char)
         (send ,pane :mouse-standard-blinker)))))

(define-grapher-command (com-move-node :menu-accelerator "Move")
    ((node 'node :prompt "click on a node" :confirm t))
    (let ((pane (send dw:*program-frame* :get-pane 'graph)))
       (with-mouse-tracking-node (node pane)
          (dw:tracking-mouse ()
          (:who-line-documentation-string
           () "Specify new location for node.  Right aborts")
          (:mouse-click (button x y)
           (unless (char-mouse-equal button #\mouse-right)
             (move node x y))             ; Don't move if clicked right
           (return nil))))))

(define-presentation-to-command-translator com-move-node-from-here-to-there
                                            (node :gesture move-node)
  (node)
  (cp:build-command 'com-move-node node))

(define-grapher-command (com-add-arc :menu-accelerator "Add Arc")
    ((node1 'node :prompt "a node")
     (node2 'node :prompt "another node" :confirm t))
    (if (eq node1 node2)
        (format t "~&Sorry, you can't make an arc from a node to itself.")
        (make-instance 'arc :node1 node1 :node2 node2)))

(define-grapher-command (com-add-arc-by-mouse :name "")
    ((node1 'node))
    (let ((node2 (accept 'node :prompt "click on the other node")))
       (com-add-arc node1 node2)
       (setf node1 nil)))
```

```
(define-presentation-to-command-translator com-add-arc-from-here-to-there
                              (node :gesture add-arc
                                    :documentation "Create arc from this node")
  (node)
  (cp:build-command 'com-add-arc-by-mouse node))

(define-grapher-command (com-delete-node :menu-accelerator "Delete Node")
    ((node 'node :prompt "click on the node to delete"
           :provide-default nil :confirm nil))
  (when (tv:mouse-y-or-n-p
          (format nil "Do you really want to delete node ~'IC~A~⊃?"
                      node))
    (delete-self node)))

(define-presentation-to-command-translator com-delete-node
  (node :gesture delete-node)
  (node)
  (cp:build-command 'com-delete-node node))

(define-grapher-command (com-delete-arc :menu-accelerator "Delete Arc")
    ((node1 'node :prompt "click on one node")
     (node2 'node :prompt "click on the other node"
            :provide-default nil))
  (let ((n-deleted 0))
    (map-over-arcs (arc)
      (when (or (and (eq (arc-node1 arc) node1)
                     (eq (arc-node2 arc) node2))
                (and (eq (arc-node1 arc) node2)
                     (eq (arc-node2 arc) node1)))
        (delete-self arc)
        (incf n-deleted)))
    (when (= n-deleted 0)
      (format t "~&Didn't find an arc between~
                 ~'IC~A~⊃ and ~'IC~A~⊃"
              node1 node2))))

(define-grapher-command (com-delete-arc-by-mouse :name nil)
    ((node1 'node))
  (let ((node2 (accept 'node :prompt "click on the other node")))
    (com-delete-arc node1 node2)))

(define-presentation-to-command-translator com-delete-arc-from-here-to-there
                              (node :gesture delete-arc)
  (node)
  (cp:build-command 'com-delete-arc-by-mouse node))

(define-grapher-command (com-reset-database :menu-accelerator "Reset" :name nil)
    ()
  (when (tv:mouse-y-or-n-p
          "Do you really want to reset the entire database?")
    (setq *all-the-nodes* nil)))
```

```
(defmacro define-grapher-mouse-gestures (&rest gesture-list)
  `(progn ,@
    (loop for (gesture mouse-char) in gesture-list
          collect `(setf (dw:mouse-char-for-gesture ',gesture)
                         ',mouse-char))))

(define-grapher-mouse-gestures
  (create-node-ex-nihilo    #\mouse-left)
  (add-arc                  #\mouse-middle)
  (delete-node              #\control-mouse-left)
  (delete-arc               #\control-mouse-middle)
  (set-node-label           #\mouse-left)
  (move-node                #\shift-mouse-left))
```

7.7 Problem Set

Questions

1. Assure yourself that you understand the code.

2. Write a command which finds any nodes with no connections to other nodes and removes them from the graph.

There are a number of problems with the program as it stands. Here's your chance to improve the teacher's work.

3. Any command causes a complete redisplay. This is really unncessary, and quite unattractive, especially if there's a lot of stuff on the screen. Fix the display routines to do minimal redisplay.

4. Currently, if two notes are connected by an arc which cuts across another node, the line for the arc just runs right through the node. Fix the **present-self** method for arcs to be smart enough to go around obstacles.

5. Make the arcs mouse-sensitive, too. To do this right requires features that are not present in Genera 7.0, like non-rectangular mouse sensitive areas (which are to be implemented in a later release).

Hints

1. Play.

2. The command should loop over all the nodes, using the **delete-self** generic function on any which have no arcs.

3. Make the redisplay for the **graph** pane be incremental. Modify the redisplay function to use the incremental redisplay functions.

4. Open problem. I haven't thought of a good way to do this.

5. Uncomment the presentation type for **arc,** and make the **present-self** method for arcs use it.

Answers

2. Here is one way to do it:

```
;;; Command to clean up unused nodes.
(define-grapher-command (com-gc-nodes
                         :menu-accelerator "GC")
      ()
  (map-over-nodes (node)
                  (when (null (node-arcs node))
                    (delete-self node)))))
```

3. One way to do it is to change **:incremental-redisplay** from **nil** to **t** in the **dw:define-program-framework** form, and then change **display-graph** to use the redisplay technology.

Here is a sketch of what's needed:

```
(defvar *tick* 0)
(defun tick () (incf *tick*))
```

```
;;; The flavor of each node in the graph.
(defflavor node
      (
       [...]
       (update-tick (tick))) ;For incremental redisplay
      ()
  (:initable-instance-variables xpos ypos label shape)
  (:writable-instance-variables label shape)
  (:readable-instance-variables arcs xpos ypos radius
                                update-tick)
  (:required-init-keywords :xpos :ypos))
```

```
(defmethod (do-tick node) ()
```

```
(setf update-tick (tick))
(loop for arc in arcs
      do (do-tick arc)))
```

Similarly, you need an **update-tick** instance variable and a **do-tick** method for arcs.

```
(defmethod ((setf node-label) node :after) (ignore)
  (do-tick self)
  (setf radius :needs-calculation))
```

```
(defmethod (move node) (new-xpos new-ypos)
  (do-tick self)
  (setf xpos new-xpos ypos new-ypos))
```

```
(defmethod (display-graph grapher) (display-pane)
  (map-over-nodes (node)
    (dw:with-redisplayable-output
      (:stream display-pane :unique-id node
               :cache-value (node-update-tick node))
      (present-self node display-pane)))
  (map-over-arcs (arc)
    (dw:with-redisplayable-output
      (:stream display-pane :unique-id arc
               :cache-value (arc-update-tick arc))
      (present-self arc display-pane))))
```

Problems 3, 4 and 5 are left for the interested reader to finish.

8. Streams and Files

In this chapter I will talk about program input/output. As with everything else on the Lisp Machine, I/O has been made as generic as possible. This means that:

- I/O is as *device-independent* as possible; the programmer need not know in advance what kind of device is actually the source of its input or the destination of its output; and

- File I/O is as *system-independent* as possible; if the user wants to use files on a Lisp Machine, or a VAX running VMS, or a machine running Unix, the programmer doesn't need to know in advance which system will actually be used.

Device-independent I/O is performed using a data abstraction called a *stream*. Regardless of what kind of device is actually involved, a stream presents a uniform interface to the programmer. This means, for example, that a program need not have to be written differently if it takes its input from a tape, a disk file, or from the console keyboard.

Similarly, system-independent file I/O involves making streams which refer to files. The way to specify *which* file you mean is with another data abstraction called a *pathname*. Each pathname refers to a "place" in a file system. A program can be written to deal with pathnames, rather than with, say, a Unix file system.

8.1 Streams

Since the mechanics of interacting with different kinds of peripheral devices vary widely, and are often quite messy, it is desirable to shield programmers from having to know the details of such operations. This shielding is accomplished by routing all input and output operations through *streams*.

A stream is an object which obeys the stream protocol. Most streams are implemented using Flavors. However, since the stream protocol is older than the current Flavors system, much of the low-level stream protocol is defined in terms of message-passing. Most of the interfaces to streams are higher-level ones which actually send the messages for you, so you might never know how the stream is implemented.

All streams accept generic "commands," to perform certain operations, like, for example, "give me the next input character," and take care themselves of the details of performing that operation on their particular sort of device. This way, knowledge about how to perform I/O operations is segregated into the stream implementations themselves, freeing programs (and programmers) from the need to understand the details of these operations. All a program needs to know is how to deal with streams;[1] the streams take care of the details.

Streams can be divided into two broad categories, depending on which kinds of Lisp objects they accept or deliver. *Character* streams operate on Lisp character objects; when asked for the next input object, they return a character. *Binary* streams operate on Lisp integers; when asked for the next input object, they return a number. Similarly, when you want to deliver output to a stream, you must pass in the correct data type: character streams want characters, and binary streams want numbers.

8.1.1 Standard Stream Operations

Some streams only handle input; some only handle output; and some do both. There is a small set of operations which all output streams are required to handle. Similarly, there is another set that all input streams are required to handle. For example, all input streams handle the :tyi operation, which returns the next input object.

Additionally, there is a somewhat larger set that all streams are guaranteed to accept, even though they themselves do not implement them directly. This bit of magic works through the *default handler*. When a stream is asked to perform an operation it does not handle directly, it invokes the default handler, which uses some combination of natively-implemented operations to produce the desired effect. For example, the default handler for character streams implements the :line-in in terms of :tyi. Some streams implement :line-in directly as well.

Most of the time, you will not need to use the low-level operations that the stream itself inherently supports. Instead, you

---

[1] And how to make the streams it wants to use. See the section "Making Other I/O Streams," page 198.

will use Common Lisp operations and Symbolics' enhancements to that set. For example, you will usually not use the **:tyi** or **:line-in** messages to a stream; instead, you will use **read-char** (or **read-byte** for binary streams) and **read-line** functions.

Here are some operations you might use on input streams of any kind:

- **read-char** – Returns the next character.

- **read-byte** – Returns the next byte (binary streams).

- **unread-char** – Puts the character back into the input stream. You may only do this with the character you just read.

- **peek-char** – Equivalent to **read-char** followed by **unread-char**; the character is returned.

- **listen** – Returns **t** if the stream contains input.

- **read-char-no-hang** – Returns the next input character, if any; returns **nil** if none.

- **read-line** – Returns the next line of input.

- **read** – Returns the next Lisp expression.

- **clear-input** – Clears any buffered input in the stream.

Here are some operations you might use on output streams of any kind:

- **write-char** – Outputs a character.

- **write-byte** – Outputs a byte to a binary stream.

- **write-string** – Writes a string of characters.

- **write-line** – Writes a string of characters, followed by a newline character.

- **terpri** – Puts out just a newline character (an old Lisp function name, apparently means "Terminate Print Line.")

- **fresh-line** – Performs a **terpri** if the current line has any characters on it. Not all streams support finding out whether they are at the left margin; if the stream does not support it, **fresh-line** *always* performs the **terpri**.

- **finish-output** – If the output stream is buffered or asynchronous, this function attempts to ensure that all output has reached its destination. This function will return only when the device has acknowledged that it has received its output.

- **force-output** – If the stream has any output buffered, it must send it to the device immediately. This function does not attempt to wait until the output actually arrives.

- **clear-output** – If there is any buffered output, it is discarded immediately without attempting to transmit it to the device.

You might also be interested in looking up the documentation for these stream operations: **input-stream-p**, **output-stream-p**, **stream-element-type**, and **close**.

8.1.2 Special-purpose Operations

There are a wide variety of operations particular to streams for one or another of the "peripherals" (files, network streams, windows, etc.). Most of these are not handled by the default handler and will result in an error if invoked on a stream which does not handle them. These are, by and large, implemented using message-passing only, since they are usually at a lower level than the other interfaces we have discussed thus far. Most of the special-purpose operations are documented along with the type of device they're intended to be used with. Here are a few of the more commonly-used of these operations:

- **:read-cursorpos** – This operation is supported by windows and other interactive streams. It returns two values, the current x and y coordinates of the cursor. Its optional argument is a keyword indicating in what units x and y should be expressed. The keywords **:pixel** and **:character** are understood by most streams.

- **:set-cursorpos** – This operation is supported by streams which support **:read-cursorpos**. It sets the position of the cursor.

- **:clear-window** – This operation erases the screen area on which this stream is displayed.

- **:clear-history** – This operation is only accepted by *Dynamic* windows; it clears *all* of the window's output, whether currently on display or scrolled off the edges.

- **:read-location** – This is a "pointer" into a file stream. It returns a value which can be used with **:set-location**. This value will always be useful to return the file to the current position, even if you close the file and re-open it, and even if you cold boot the machine, having saved this location away someplace where you can get it back.[2]

---

[2]Some binary streams implement an even more primitive version of this message, **:read-pointer**, instead of **:read-location**.

- **:set-location** – Sets the stream's "pointer" so the next place you read or write to it is the same as it was when you obtained the location using **:read-location**.

8.1.3 Standard Streams

There are a number of special variables whose values are streams widely used by many system (as well as user) functions. Here are some of them:

standard-input In the normal Lisp top-level lop, input is read from ***standard-input*** (*i.e.*, whatever stream is the value of ***standard-input***.) Many input functions, including **read** and **read-char**, take a stream argument which defaults to the value of ***standard-input***. Most of your input should come from ***standard-input***.

standard-output

Analogous to ***standard-input***; in the Lisp top-level loop, output is sent to the stream which is the value of ***standard-output***, and many output functions, including **write-char** and **print**, take a stream argument which defaults to ***standard-output***. Most of your output should be sent to ***standard-output***.

terminal-io The value of ***terminal-io*** is the stream which connects directly to the user, normally through the system console. In a process which is associated with a window, the value of ***terminal-io*** is likely to be that window. For background processes, for example, those created using the function **process-run-function**, ***terminal-io*** is a special "background" stream which notifies the

user when the process wants input or to do
output. This stream appears as a little background window when the user exposes it
using Function θ S. For certain other
processes (such as the keyboard process), the
value of *terminal-io* is a special marker
which will force the use of the cold load
stream if any attempt is made to use it for
I/O.

Some other useful stream variables are: *error-output*,
trace-output, *query-io* and *debug-io*, which are used for
printing errors, printing trace displays, asking questions, and
debugging, respectively.

standard-input, *standard-output* and several others and
initially bound to *synonym streams* which pass all operations on
to the stream which is the current value of *terminal-io*. So,
if *terminal-io* is re-bound, the synonym streams see the new
value.

User programs generally don't change the value of
terminal-io. A program which wants, for example, to divert
the output to a file should do so by temporarily binding
standard-output; that way, error messages sent to
error-output can still get to the user by going through
terminal-io, which is usually what is desired.

8.2 Accessing Files and Directories

Some of the information in this section will make more sense
after reading the following section.

8.2.1 Open and Other Functions for Operating on Files

All reading from and writing to a file is done through streams. To access a file, you must have an open stream to that file. The fundamental way to obtain an open stream to a file, whether for reading or writing, is with the **open** function.

open   Takes a pathname and a sequence of keyword options, and returns a stream connected to the specified file. The pathname may be anything acceptable to **fs:parse-pathname**, generally either a string or an actual pathname object. Some of the more frequently used **open** keywords include **:direction**, which specifies whether the file is to be read or written, **:element-type**, which specifies whether you want a binary or character stream, and **:if-does-not-exist**, which tells **open** what to do if the file is not there.

Most programs do not call **open** directly. They more commonly use the **with-open-file** macro, which makes use of an **unwind-protect** to guarantee that the stream will be closed when you're done with it. If you call **open** directly, you should also use an **unwind-protect** to make sure the stream gets closed, because leaving around lots of open streams can create problems.

with-open-file   Evaluates the *body* forms with the variable *stream* bound to a stream opened by applying **open** to the pathname and option list. When control leaves the body, either normally or abnormally, the file is closed. If control leaves abnormally (*i.e.*, because of an error or a **throw**), the file is closed in *abort* mode; output files closed in abort mode are deleted.

So, if I wanted to write a new text file in my directory on my Lisp Machine file server Cerridwyn, which contained only the string "Wow, I'm on a disk!" (without the double quotes), I would evaluate:

```
(with-open-file (stream "CD:>rsl>yippee" :direction :output)
  (write-string "Wow, I'm on a disk!" stream))
```

Or, if I wanted to see how many characters into a certain file the first "a" occurred,

```
(with-open-file (stream "CD:>rsl>lispm-init.lisp")
  (loop for i upfrom 1
        as char = (read-char stream)
        when (char= char #\a) return i))³
```

Here are some more functions for operating on files. They generally accept either a string or a pathname object, and some of them also accept a stream open to the appropriate file.

rename-file   Changes the name of the file. The CP command `Rename File` and the editor command `Rename File` (m-X) use this function. See the documentation for details on what happens with links.

delete-file   The specified file is deleted.

copy-file   Copies one file to another. The CP command

---

³Watch out! **char=** compares for an exact match. Not only won't it return t for (capital) A, it won't return t for a lower-case A which appears in the file in some other character style. See the section "Overview of Characters" in *Symbolics Common Lisp: Language Concepts*.

Copy File and the editor command Copy File (m-X) use this function. See the documentation for details of merging, wildcard names, and links, and for the meanings of its keyword arguments.

probe-file If the specified file exists, returns the pathname of its "true name", *i.e.,* its ultimate destination in the file system. Note that this might not be the same as the argument; in the presence of links it might be wildly different.

load Loads the specified file into the Lisp environment. If it's a text file, it is evaluated; if a binary file, it's loaded using the binary ("fasload") loader.

fs:file-properties returns a list: the first element of the list is the "truename"; the rest of the list is a list of alternating keyword/value pairs, suitable for use with **getf**. The second value returned by **fs:file-properties** is a list of those properties which can be set using **fs:change-file-properties**. See the documentation for a list of the possible file property indicators.

fs:change-file-properties

takes a pathname, an error-p flag, and a list of alternating keyword/value pairs. **fs:change-file-properties** alters the attributes of the file accordingly, if possible. (Some properties are not alterable. Which ones are is a property of the host file system and the user's privileges on that file system.)

Certain operations are defined on file streams:

pathname   returns the pathname of the file to which the
                        stream is attached. [**pathname** also works
                        on strings and pathname objects, returning
                        pathname objects in all cases.]

file-length   returns the number of elements in the file.
                        For a binary file, the length is in units of the
                        :element-type specified when the file is
                        opened.

8.2.2 Directories

directory   returns a list of files which matches the
                        given pathname.

fs:directory-list   finds all files matching the pathname, and
                        for each one gets the information that would
                        be returned by **fs:file-properties** for that file.
                        It collects all of these into a list, and adds
                        one element to the list containing information
                        about the file system as a whole; this added
                        element has **nil** in the place where the path-
                        name goes.   The returned list has one more
                        element than there are files matching the
                        pathname.   See the function **fs:directory-list**
                        in *Reference Guide to Streams, Files, and I/O*.

8.3 Pathnames

Just as streams are intended to provide a uniform, device-
independent interface between programs and the different kinds
of peripherals, *pathnames* are intended to provide a uniform in-

terface between programs and remote files systems. The idea is to free the programmer from having to keep in mind the format for file names on the various remote hosts. With pathnames, you should be able to manipulate files on a file server without knowing anything about that server's syntax for file names.

All pathnames are instances of some flavor built on **pathname**. Each pathname has six components which correspond to different parts of a file name. The mapping of the components into the parts of the file names is done by the pathname software, and is specific to each kind of host the software knows about.

The six components of a pathname are the *host*, the *device*, the *directory*, the *name*, the *type*, and the *version*. So, for example, the pathname corresponding to the string "CD:>rsl> lispm-init.lisp.105" might be an object of flavor **fs:lmfs-pathname:**[4]

```
#P"CD:>rsl>lispm-init.lisp.105",
        an object of flavor FS:LMFS-PATHNAME,
  has instance variable values:
    FS:HOST:              #<LISPM-HOST CERRIDWYN 700106>
    FS:DEVICE:            :UNSPECIFIC
    FS:DIRECTORY:         ("rsl")
    FS:NAME:              "lispm-init"
    FS:TYPE:              "lisp"
    FS:VERSION:           105
```

The pathname for the file /usr/hjb/lispm-init.l on (4.2bsd) Unix host "sola" would look like this:

---

[4]Pathnames actually contain more instance variables than this. The others are for caching certain strings that represent the pathname in different contexts.

```
#P"SOLA:/usr/hjb/lispm-init.l",
        an object of flavor FS:UNIX42-PATHNAME,
    has instance variable values:
    FS:HOST:                    #<UNIX-HOST SOLA 707345>
    FS:DEVICE:                  :UNSPECIFIC
    FS:DIRECTORY:               ("usr" "hjb")
    FS:NAME:                    "lispm-init"
    FS:TYPE:                    "l"
    FS:VERSION:                 :UNSPECIFIC
```

The special keyword **:unspecific** is used as a place-holder for components which are not filled in; **nil** is used when the user intentionally omits the component.

A pathname need not refer to a specific file. #P"CD:" is a perfectly legitimate pathname, even though it specifies only a host and nothing else.

The conversion of a string into a pathname is done with the function **pathname**.[5] The first thing it has to do is determine the host, since the methods for parsing the rest of the components depends on which host it is. If there are any colons in the input string, everything appearing before the first colon is considered to be the name of the host.[6] Parsing of the remainder proceeds according to the type of the host, and its own syntax for file names. (If there are no colons, some default value is used for the host – every pathname *must* have a host.)

---

[5] Older code might use the function **fs:parse-pathname**.

[6] If the colon is the first character in the string, it is as if no host were specified, and the default is used; this is useful for those hosts whose native syntax contains colons, such as VMS, which uses colons to delimit disk names.

Of course, there's no need to go through strings (and worry about the remove hosts's file name syntax) at all. One of the selling points of pathnames is precisely that you shouldn't need to do so. Accordingly, one may construct pathnames in this manner:

```
(make-pathname :host "CD" :directory '("RSL")
               :name "LISPM-INIT" :type "LISP"
               :version 105)
```

which returns the same pathname whose printed representation was shown above.

Pathnames are *interned*, just like symbols, meaning that there is never more than one pathname with the same set of component values. The main reason for maintaining uniqueness among pathnames is that they have property lists, and it's desirable for two pathnames that look the same to have the same property lists.[7] Thus, the **make-pathname** expression just above returns the same (**eq**) pathname object each time.

8.3.1 Component Values

The *host* component is always a host object (an instance of some flavor built on **net:basic-host**). The permissible values for the other components depends to some extent on the type of the host, but there are some general conventions.

The *type* is always a string, or one of the symbols **nil**, **:unspecific**, or **:wild**. Both **nil** and **:unspecific** denote a miss-

---

[7]Sometimes there are clusters of files which go together, such as a source and object file for the same program. In order to share properties between these clusters, a single pathname representing the entire cluster is chosen. This pathname is called the *generic pathname*, and is derived by sending the **:generic-pathname** message to the pathname representing any member of the cluster.

ing component. The difference is in what happens during *merging* (see below); **nil** generally means to use the default, and **:unspecific** means to keep that component empty. The symbol **:wild** is sometimes used in pathnames given to **directory**, and matches all possible values.

The type field gives an indication of what sort of stuff is in the file. Lisp source files, for instance, usually have a type component of "lisp," and compiled Lisp code a type component of "bin." Since there are some system-dependent restrictions on how many characters may appear in this field, a *canonical type* mechanism exists to allow processing of file types in a system-independent fashion.

The Symbolics documentation says: "A *canonical type* for a pathname is a symbol that indicates the nature of a file's contents. To compare the types of two files, particularly when they could be on different kinds of hosts, you compare their canonical types." For instance, a Lisp source file on a VMS system might have a file type of 'LSP,' and one on a UNIX system might have a file type of 'l.' When we ask these pathnames for their canonical type, we receive the keyword symbol **:lisp**.

The *version* is either a number or one of the symbols **nil**, **:unspecific, :wild, :newest** or **:oldest**. The first three have the same meaning as for the type component. **:newest** refers to the largest number that exists (when reading a file) or one greater than that number (when writing a new file). **:oldest** refers to the smallest number that exists.

The *device* component may be either **nil** or **:unspecific**, or a string designating some device, for those file systems that support such a notion (VMS, TOPS-20, ITS).

The *name* component may be **nil, :wild** or a string.

The *directory* component may be **nil** or **:wild** for any type of host. On non-hierarchical file systems, a string is used to specify a particular directory. On hierarchical systems, the directory component (when not **nil** or **:wild**) can be the keyword **:root** or a list of *directory level components*. These are usually strings. So the pathname #P"CD:>rsl>text>network.text" has for its directory component the list ("RSL" "TEXT").

The directory level components can also be special keywords as well as strings. **:wild** matches any single directory. **:wild-inferiors** matches zero or more directory levels. So, the pathname #P"SOLA:/usr/*/lispm-init.l" has ("USR" :wild) as its directory component, and the pathname #P"CD:>rsl>**> examples>*.lisp.newest" has ("RSL" :wild-inferiors "EXAMPLES") for its directory component. Directory level strings may also be partially wild, like "*FOO*".

The keyword **:relative** may appear first, meaning that the pathname doesn't start at the root (such a pathname can only be valid for describing actual files when merged with an absolute pathname). **:relative** may be followed by zero or more of the symbol **:up**, which means that when it is merged with a pathname, it first deletes that many components from the other pathname's directory component. Thus, the pathname #P"SOLA:../foo/bar" has (:relative :up "FOO") for its directory component, and #P"CD:<<text>new>patch.lisp" has a directory component of (:relative :up :up "TEXT" "NEW").

8.3.2 Case in Pathnames

Since the various host systems have different conventions as to upper and lower case characters in file names, most pathname functions perform some standardization of case to facilitate manipulating pathnames in a host-independent manner. There are two representations for any given component value, one in *raw* case and one in *interchange* case. Raw case representation, which is used internally for the instance variables of pathnames, corresponds exactly to what would be sent to the remote machine to find the file corresponding to the pathname. Interchange case is the standardized form, and is what you get if you ask a pathname for its component values. It's also what functions like **make-pathname** expect.

The standardization is simple. Each type of host is classified as to whether its preferred case, or *system default case*, is upper or lower. Any raw component which is in the preferred case for its host has an upper case interchange form. A raw component in mixes case has an identical (mixed case) interchange form. Since LMFS and UNIX hosts are classified as having lower case for the system default, this means that the raw forms are case-inverted to get the interchange forms, and *vice versa*.

The normal messages for accessing and setting the component values of pathnames are based on interchange form. There are other messages which deal explictly with raw form. Note that the pathname parsing routines deal with raw form, since they expect you to type pathnames in the form in which the host expects them.[8]

8.3.3 Defaults and Merging

In most situations where the user is expected to type in a pathname, some *default pathname* is displayed, from which the values of the components not specified by the user may be taken. Many programs maintain their own default pathnames,

---

[8]If a host is an upper-case-by-default host, pathnames will be uppercased when parsed. However, it is possible to specify mixed- and lower-case components with the pathname-combining functions and messages.

containing component values that would be reasonable in the particular context. For programs which really have no idea of what sort of pathname to expect, there is a set of *default* defaults.

The pathname provided by the user (actually, the pathname constructed by **pathname** from the string provided by the user) and the default pathname are then *merged* by the fuction **merge-pathnames**. The details are a little messy, but the basic idea is that components which aren't specified in the user's pathname are taked from the default.

The programmer need not know much of this as of Genera 7.0. The **pathname** presentation type takes care of merging as well as parsing the string typed in by the user. Some variant of (accept 'pathname) is really all that is required in most cases.

8.3.4 Pathname Functions and Methods

We've already seen two of the most important pathname functions: **pathname** and **merge-pathnames**. Here are a few more:

make-pathname In addition to the keywords it tells you about, you can also specify such things as **:raw-directory**, **:raw-name** and so forth. You can also specify **:canonical-type**, which takes the canonical type as a keyword and inserts the host-specific type correctly into the new pathname.

fs:define-canonical-type

is how you define a new canonical type. The canonical-type argument is the symbol for the new type. The body is a list of specifications giving the *surface* type(s) corresponding

to this canonical type for various host types.
The default is the string used for any host
types not mentioned in the body.   Here is
how the **:lisp** canonical type is defined:

```
(DEFINE-CANONICAL-TYPE :LISP "LISP"
  ((:TENEX :TOPS-20) "LISP" "LSP")
  (:UNIX "L" "LISP")
  (:UNIX42 "LISP" "L")
  (:VMS4 "LISP" "LSP")
  ((:VMS :MSDOS) "LSP"))
```

Like the window system, the pathname system predates the current implementation of Flavors, and much of the interface to
them is in terms of messages.

- The **:host** message to pathnames returns the host component, which is an instance of some host flavor.

- The messages **:device**, **:directory**, **:name** and **:type**
  return the corresponding component value, with any
  strings given in interchange case.

- The messages **:raw-device**, **:raw-directory**, **:raw-name**
  and **:raw-type** are similar, but use raw case for all
  strings.

- The **:version** message returns the version (case is not an
  issue, since versions are never strings).

- The **:canonical-type** message returns two values;
  together, they indicate the type component of the pathname, and what canonical type – if any – it corresponds
  to (See the documentation for details).

- The messages **:new-host**, **:new-device**, **:new-directory**,

:new-name and :new-type all take one argument and return a new pathname which is just like the one that received the argument except that the value of the specified component will be changed. All strings will be accepted in interchange case (except for hosts, which are converted to host objects).

- You can probably guess what :new-raw-device, :new-raw-directory, :new-raw-name and :new-raw-type do.

- :new-version and :new-canonical-type do the obvious thing, and have no "raw" form for obvious reasons.

- :new-pathname allows wholesale replacement of component values. Its arguments are alternating keywords and values, with the same keywords accepted as by make-pathname.

- There are a set of messages for getting strings that describe the pathname. The returned strings come in different forms for different purposes. :string-for-printing returns the string that you see inside the printed representation of a pathname. :string-for-host shows the file name (not including the host) the way the host file system likes to see it. There are several others.

- :get, :putprop, :remprop and :plist all do the obvious thing with the pathname's property list. Keep in mind the distinction between the pathname's property list and the list returned by fs:file-properties. The latter are properties of a file, and require accessing the host's file system. The former are the properties of a pathname, a Lisp object which may not even correspond to any files.

Directory pathnames are a special case in most hierarchical file systems. There are times when you would like to know about the "location" of the directory in its file system, and other times when you really want to manipulate the directory's contents. For example, to rename a directory, you need to know the former pathname, while to open a file in the directory you want the latter.

Two methods have been defined on pathnames which converts one form of pathname to the other:

- **:pathname-as-directory** converts "location" pathnames into "contents" pathnames.
- **:directory-pathname-as-file** performs the inverse operation.

This is best explained by example:

```
(send #P"CD:>rsl>foo.directory" :pathname-as-directory)
  -> #P"CD:>rsl>foo>"

(send #P"CD:>rsl>foo>" :directory-pathname-as-file)
  -> #P"CD:>rsl>foo.directory.1"
```

Here is a small sample program which lists the files in all my top-level subdirectories:

```
(defun list-all-subdirs (&optional (path (fs:user-homedir)))
  (setq path (send path :new-pathname
                    :name :wild
                    :type :wild
                    :version :wild))
  (loop for (path . props) in (fs:directory-list path)
        when (and path
                  (getf props :directory))
          append (directory
                   (send (send path :pathname-as-directory)
                         :new-pathname
                         :name :wild
                         :type :wild
                         :version :wild))))
```

8.3.5 Logical Pathnames

There are some pathnames which don't correspond to any particular file server, but rather to files on a *logical*[9] *host*. The logical host may then be mapped onto any actual host, thus defining a translation from "logical pathnames" to "physical pathnames." This feature improves the transportability of code.

Take the Lisp Machine software as an example. Every Lisp Machine site keeps the source code on a different computer. There are many functions that want to be able to find these files, no matter what site they're running at. The solution is to use logical pathnames: all the system software is in files on the logical host "SYS". Each site gives the "SYS" host an appropriate definition, and then it works just to find open a file with a name like "SYS:IO;PATHNM.LISP", which happens to be the file containing the bulk of the pathname code. At my site

---

[9]See hacker's definition at the end of chapter.

that corresponds to the file "QX:>sys>io>pathnm.lisp;" at your site it might be the VMS pathname "GENIE:SYS$SYMB: [REL-7.IO]PATHNM.LISP."

The function **fs:set-logical-pathname-host** defines the mapping of file names from a logical host to the corresponding physical pathnames. Rather than embedding this call in your code (remember, the idea of this is transportation of code from site to site without modification), you put it in the site directory in a file called "SYS:SITE;<*host-name*>.TRANSLATIONS." Then, if you call the function **fs:make-logical-pathname-host** with an argument of the host name, it will look for and load the appropriate file, thus evaluating the **fs:set-logical-pathname-host**.

The use of **fs:set-logical-pathname-host** is best explained by example. Figure 5 contains abridged version of the contents of "SYS:SITE;SYS.TRANSLATIONS," which defines the location of the Symbolics-supplied files at my site.

As you can see, this file contains Lisp code, and is read by the normal Lisp reader. Some notes:

- This file is read and evaluated using the normal Lisp file loader. It contains *Lisp Code*. That means you can use conditionals, reader macros, etc., to make it do what you want. It also means that it can contain anything you like, as long as it eventually uses **fs:set-logical-pathname-host** someplace.

- This particular file has a conditional at the top level; it is dispatching on the release of Lisp Machine software running on the reading machine. For each of releases 6 and 7, it does something different. It signals an error if you try to read it into some other release.

- Each of the branches of the dispatch invokes

```
;;; -*- Mode: Lisp; Syntax: Common-Lisp; Base: 10; -*-

(ecase (si:get-release-version)

;;; Release 6
  ((6)
   (fs:set-logical-pathname-host
    "SYS"
    :no-translate t
    :translations '(("SYS: **;" "CD:>sys>**>"))))

;;; Release 7
  ((7)
   (fs:set-logical-pathname-host
    "SYS"
    :physical-host "CERRIDWYN"
    :translations
    '(
      ;; Release-independent files
      ("SYS:FONTS;**;*.*.*" "CD:>sys>fonts>**>*.*.*")
      ("SYS:L-UCODE;**;*.*.*" "CD:>sys>l-ucode>**>*.*.*")
      ("SYS:DATA;**;*.*.*" "CD:>sys>data>**>*.*.*")
      ("SYS:N-FEP;**;*.*.*" "CD:>sys>N-Fep>**>*.*.*")
      ;; Everything else.
      ("SYS:**;*.*.*" "QX:>sys>**>*.*.*")))))
```

Figure 5.   A Sample SYS:SITE;SYS.TRANSLATIONS

fs:set-logical-pathname-host.   The release-6 version is very straightforward; it merely says to find all files on the SYS logical host in corresponding places under CD:>sys>. The release-7 one is slightly more confusing. It says that *certain* directories on the SYS logical host are found on CD, in the same places as they are found under release-6 (these files are those which do not change between releases, such as microcode files and fonts.) Everything else is found in the corresponding directory on *another* physical host, namely the Lisp machine named QX.

- As must be obvious, logical hosts do not necessarily correspond to single physical hosts. You can spread the files out to as many different hosts as you like. They needn't even be the same *type* of host: you could choose to keep some files on, say, a Lisp Machine, and others on a VMS, Unix, TOPS-20 or Multics host.

Given a pathname for some logical host, the mapping to physical pathname is carried out by sending the logical pathname the **:translated-pathname** message (you can send this to physical pathnames as well; they merely return themselves). So, for example, under release 7, the file which defines pathnames (`#P"SYS:IO;PATHNM.LISP"`) is translated into `#P"QX:>sys>io>pathnm.lisp"` at my site; the font file `#P"SYS:FONTS;TV;CPTFONT.BFD"` becomes `#P"CD:>sys>fonts>tv>cptfont.bfd"`.

8.4 Making Other I/O Streams

Earlier, we showed you how file streams are made, using the function **open**. Here are some other stream-making functions to get you started hacking:

- **hardcopy:make-hardcopy-stream** – a stream whose output is printed on a printer.

- **tape:make-stream** – a tape stream, supporting either industry standard 9-track tapes or 1/4-inch cartridge tapes.

- **si:make-serial-stream** – an RS-232 port, supporting asynchronous devices up to 19,200 baud.

- **tv:make-window** – a window on the screen.

- **zwei:open-editor-stream** – a stream which is attached to an editor buffer.

Common Lisp provides a number of primitives for creating streams:

- **make-concatenated-stream** – used to make a single input stream out of a set of streams. Taking input from this stream in turn takes input from the underlying streams until each is exhausted. When the last one is at end-of-file, the concatenated stream returns EOF.

- **make-broadcast-stream** – used to make a single output stream which broadcasts to several underlying streams.

- **make-string-input-stream** – makes a stream which takes its input from the string provided.

- **make-string-output-stream** – makes a stream which places its output into a string.

8.5 Fun and Games

From *The Hacker's Dictionary*, Guy L. Steele, Jr., *et al*:

LOGICAL *adjective.*
Conventional; assumed for the sake of exposition or convenience; not the actual thing but in some sense equivalent to it; not necessarily corresponding to reality.

Example: If a person who had long held a certain post (for example, Les Earnest at Stanford) left and was replaced, the replacement would for a while be known as the "logical

Les Earnest." Pepsi might be referred to as "logical Coke" (or vice versa).

At Stanford, "logical" compass directions denote a coordinate system in which "logical north" is toward San Francisco, "logical south" is toward San Jose, "logical west" is toward the ocean, and "logical east" is away from the ocean – even though logical north varies between physical (true) north near San Francisco and physical west near San Jose. The best rule of thumb here is that El Camino Real by definition always runs logical north-and-south. In giving directions, one might way, "To get to Rincon Tarasco Restaurant, get onto EL CAMINO BIGNUM going logical north." Using the word "logical" helps to prevent the recipient from worrying about the fact that the sun is setting almost directly in front of him as he travels "north."

A similar situation exists at MIT. Route 128 (famous for the electronics industries that have grown up along it) is a three-quarters circle surrounding Boston at a radius of ten miles, terminating at the coast line at each end. It would be most precise to describe the two directions along this highway as being "clockwise" and "counterclockwise," but the road signs all say "north" and "south," respectively. A hacker would describe these directions as "logical north" and "logical south," to indicate that they are conventional directions not corresponding to the usual convention for those words. (If you went logical south along the entire length of Route 128, you would start out going northwest, curve around to the south, and finish headed due east!)

8.6 Problem Set

Questions

1. Write a function which prints an arbitrary **format** string into a file in your home directory; it should take the name of the file as an argument, and the type of the file should be "text." The rest of its arguments should be just like **format**: control string and optional arguments.

2. Write a similar function which does the same thing, only it appends the new output to whatever is already in the file.

3. Add a form to your init file which records all your logins; it should put the time and the fact that you've logged in in the file "login-history.text" in your home directory.

4. *(Requires more research)* Add a form to your init file which records all your logouts. It should put it in the same file that your login history goes in.

5. Suppose there is a binary file whose contents are numbers in the range $-2,147,483,648 \leq n \leq 2,147,483,647$ (32-bit integers). How can we read the file into an array of fixnums? It'd be convenient to open a 32-bit binary stream to the file and just do **:tyi**'s or a **:string-in**, but most file servers won't allow a 32-bit stream. We'll have to use a 16-bit stream. One strategy is to read two 16-bit bytes at a time and build a 32-bit number by shifting one number 16 bits and adding them together. This will work, but it's awfully slow. Can you think of anything better? (Hint: think about displaced arrays of different types.)

Answers

1. The trick here is to use the Lisp function **apply** to pass a
&rest argument to **format**:

```
(defun format-into-file (name format-string
                              &rest format-args)
  (with-open-file (file (send (fs:user-homedir)
                              :new-pathname
                              :raw-name name
                              :canonical-type :text)
                        :direction :output)
    (apply #'format file format-string format-args)))
```

2. Almost the same, but you need to specify the **:direction**
and **:if-does-not-exist** keyword arguments.

```
(defun format-append-file (name format-string
                                &rest format-args)
  (with-open-file (file (send (fs:user-homedir)
                              :new-pathname
                              :raw-name name
                              :canonical-type :text)
                        :direction :append
                        :if-does-not-exist :create)
    (apply #'format file format-string format-args)))
```

3. Place this into your init file, and you have it:

```
(format-append-file "login-history"
                    "Login by ~A at ~\\datime\\~%"
                    si:*user*)
```

4. The variable **sys:logout-list** is provided for this very purpose.

```
(push '(format-append-file
        "login-history"
        "Logout by ~A at ~\\datime\\~%"
        si:*user*)
      sys:logout-list)
```

5. The slow way:

```
(defun read-file-32-slow (file &optional array)
  (with-open-file (stream file :element-type
                              '(unsigned-byte 16))
    (unless array
      (setq array
            (make-array (floor (send stream :length)
                               2))))
    (loop for i from 0
          for c1 = (read-byte stream nil)
          for c2 = (read-byte stream nil)
          while c1
          do (setf (aref array i) (+ c1 (lsh c2 16))))
    array))
```

The fast way:

```
(defun read-file-32-fast (file &optional array32)
  (with-open-file (stream file :element-type
                          '(unsigned-byte 16))
    (unless array32
      (setq array32
            (make-array (floor (send stream :length) 2)
                        :initial-value 0)))
    (send stream :string-in nil
          (make-array (* 2 (length array32))
                      :element-type '(unsigned-byte 16)
                      :displaced-to array32))
    array32))
```

9. The Calculator Example

This chapter is very much like the graph example two chapters back – a later section contains a code listing, and this one describes some of the new features[1] of the system used in the code. This program is derived from a program originally written by Dan Weinreb, and rewritten for Genera 7.0 by Mike McMahon.

Once again, if your site has the tape for this book, you can load the code by using the CP command Load System Calculator. After the code has been loaded, start the program by typing Select +. The calculator frame will look something like figure 6.

9.1 The Program Frame

The calculator program simulates a hand-held calculator doing "reverse Polish notation" (RPN) calculations. Thus, you would

---

[1] See hacker's definition at end of chapter.

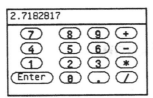

Figure 6.   Calculator program display window

key in the two arguments of an operator first, separating them with Enter, as required, followed by the operator itself. In this program, the "keyboard" is a menu: you can click on various "buttons" in the display.   Also, the regular console keyboard can be used to type in the same commands: each digit you type is as if you had "pressed" (*i.e.*, clicked on) the equivalent button; the other commands are as you see them, except that Enter is typed with the Return key.

As in the previous example, the calculator program framework was written with "Frame-Up" (Select Q) and then transferred to the editor. Thus, most of it was written automatically.

The essential differences between this version and the last are:

- The command table includes keyboard accelerators. This is how the one-character commands work. If you type "3.14159" as input, it is exactly equivalent to clicking on those characters in the calculator's keyboard pane.

- The top-level is specified. This is not because we want to specify the top-level function (we are using the default one, after all), but because we want to turn off echoing. You may recall that every time you clicked on a node for setting its label, or moving it, etc., that the command was echoed in the interaction pane. This is how to turn it off.

- The value pane uses incremental redisplay. More about this in the next section. You can also see how to specify the default character style for a pane in this description.

- The keyboard pane specifies **:more-p nil**. Otherwise, when its redisplay function tries to write the bottom line of the display, it might get stuck in a "****more****" break.

- This program definition uses the **:size-from-pane** option. This is how the calculator winds up being only a couple of inches on a side, rather than taking up the entire screen.

9.2 The Redisplay

Using incremental redisplay is completely simple in this example. The display has only one item which might need to be updated. To do this, the program uses the function **dw:redisplayable-format**. For further information: See the function **dw:redisplayable-format** in *Programming the User Interface, Volume a.*

The keyboard layout is somewhat more complicated. This program could have used the pane-type **:command-menu**, except its author wanted more complete control over the layout of the menu:

- The command menu items aren't all in the same character style. In particular, the "Enter" key is printed using the **:large** size, while the remainder are printed using **:very-large**.

- The oval borders surrounding the items could not be displayed using the **:command-menu** pane type.

The function **dw:program-command-menu-item-list** takes a program, and optionally a menu level, and returns two values. The first is a list of command strings which need to be displayed. Its order is undefined. The second value is the presentation type which must be used when presenting the command menu items. If you present the command name using this presentation type, they will translate into commands when clicked on.

The remainder of the redisplay function is straightforward if a bit complex. The keyboard layout comes from the keyboard layout variable. The nested loops are used to display the keyboard in the order shown in that layout. The normal table layout macros are used (**formatting-table** *et. al.*), and the border is drawn with **surrounding-output-with-border**. The character size is chosen based on whether the menu item is a single character or a string.

This redisplay function could also have been written as a method for the flavor **calculator**, as was the one for the value. It didn't need access to any of the instance variables, though, so there was no reason to do so.

9.3 The Command-definition Macrology

Again, what's going on here is that we want to adopt as much of the command menu and keyboard accelerator technology as possible without having to use command menus in the default way. Here, the primary aim is to define the digit and arithmetic command function only once; it would be trivial to write a macro that created, say, ten digit commands.

The macro **define-digit-command** uses two lower-level macros which do portions of the job that **define-calculator-command** would do with its **:menu-accelerator** and **:keyboard-accelerator** options:

- **dw:define-command-menu-handler** is used to add the decimal string for the digit to the command menu at level **:top-level**. The command menu item translates into the **com-digit** command.

- **cp:define-command-accelerator** is used to add the character representing the decimal digit to the command accelerator table for the command table. It, too, translates into the **com-digit** command.

Both of these macro invocations depend on the fact that the name of the command table defined for a program is the same as the name of the program.

The macro **define-digit-commands** uses **define-digit-command** ten times, for each of the decimal digits. It is used at top level in the next form.

define-arithmetic-command works the same way as **define-digit-command**, except it translates its menu items into **com-arithmetic**.

9.4 The Program

lisp-lore:examples:calculator:calculator-system.lisp

```
;;; -*- Mode: LISP; Syntax: Common-lisp; Package: USER; Base: 10 -*-

(defpackage calculator
  (:use scl)
  (:colon-mode :external))

(defsystem calculator
    (:default-pathname "lisp-lore:examples;calculator;"
     :maintaining-sites :ssf
     :pretty-name "Calculator demo program")
  (:serial "calculator"))
```

lisp-lore:examples:calculator:calculator.lisp

```lisp
;;; -*- Mode: LISP; Syntax: Common-lisp; Package: CALCULATOR; Base: 10; Lowercase: Yes -*-

;;; The program framework
(dw:define-program-framework calculator
  :select-key #\+
  :command-definer t
  :command-table (:inherit-from nil :kbd-accelerator-p t)
  :top-level (dw:default-command-top-level :echo-stream ignore)
  :panes ((value :display :redisplay-function 'calculator-display-value
                          :incremental-redisplay t
                          :default-style '(:fix :roman :large)
                          :height-in-lines 1)
          (keyboard :display :redisplay-function 'calculator-draw-keyboard
                             :redisplay-after-commands nil
                             :more-p nil))
  :label-pane nil
  :size-from-pane keyboard
  :state-variables ((current-value 0.0)          ;The current value displayed.
                    (value-stack nil)            ;The stack of pushed values.
                    (entry-state 'new)           ;State controlling meaning of digits.
                    )
  )

;;; The redisplay method for the value pane
(defmethod (calculator-display-value calculator) (stream)
  (dw:redisplayable-format stream "~F" current-value))

;;; The layout of the calculator keyboard.
(defvar *calc-keyboard-layout* '(("7" "8" "9" "+")
                                 ("4" "5" "6" "-")
                                 ("1" "2" "3" "*")
                                 ("Enter" "0" "." "/")))

;;; The redisplay method for the keyboard pane
(defun calculator-draw-keyboard (program stream)
  (multiple-value-bind (item-list presentation-type)
      (dw:program-command-menu-item-list program)
    (formatting-table (stream :inter-row-spacing 5 :inter-column-spacing 10)
      (dolist (sublist *calc-keyboard-layout*)
        (formatting-row (stream)
          (dolist (name sublist)
            (unless (member name item-list :test #'equal)
              (error "The item ~S was in the layout but not a defined menu item." name))
            (formatting-cell (stream :align :center)
              (dw:with-output-as-presentation (:object name :type presentation-type
                                               :stream stream
                                               :single-box t :allow-sensitive-inferiors nil)
                (surrounding-output-with-border (stream :shape :oval)
                  (with-character-size ((if (> (string-length name) 1)
                                            :large :very-large)
                                        stream
                                        :bind-line-height t)
                    (write-string name stream)))))))))))
```

```
;;; The command associated with each digit command.
(define-calculator-command (com-digit) ((value 'number))
  (cond ((eq entry-state 'new)
         ;; Digit means start building a new number.
         (push current-value value-stack)
         (setq current-value (float value)
               entry-state 'continue))
        ((eq entry-state 'new-no-push)
         ;; Digit means start building a new number but don't push.
         (setq current-value (float value)
               entry-state 'continue))
        ((eq entry-state 'continue)
         ;; Digit means continue building the number.
         (setq current-value (+ (* current-value 10.0) value)))
        ((numberp entry-state)
         ;; Digits means continue building fraction part of the number.
         (incf current-value (* entry-state value))
         (setq entry-state (* .1 entry-state)))))

;;; A helper macro for defining digit commands
(defmacro define-digit-command (num)
  `(progn
     (dw:define-command-menu-handler (,(format nil "~D" num) calculator (:top-level))
         ()
       '(com-digit ,num))
     (cp:define-command-accelerator ,(intern (format () "COM-~D" num)) calculator
                                    ,(digit-char num) () ()
       '(com-digit ,num))))

;;; The macro which actually defines all the digit commands.
(defmacro define-digit-commands ()
  `(progn . ,(loop for n from 0 to 9 collect `(define-digit-command ,n))))

;;; Do it.
(define-digit-commands)

;;; The "." command
(define-calculator-command (com-decimal-point :menu-accelerator "." :keyboard-accelerator #\.)
    ()
  (when (eq entry-state 'new)
    (push current-value value-stack))
  (when (member entry-state '(new new-no-push))
    (setq current-value 0.0))
  (when (not (numberp entry-state))
    (setq entry-state .1)))
```

```
;;; The arithmetic commands: +, -, *, /.  Depends on the command name
;;;  being the same as the function name.
(define-calculator-command (com-arithmetic)
                           ((fun 'sys:function-spec))
  (setq current-value (funcall fun (or (pop value-stack) 0.0) current-value))
  (setq entry-state 'new))

;;; A helper macro for arithmetic command definitions
(defmacro define-arithmetic-command (fun)
  `(progn
     (dw:define-command-menu-handler (,(string fun) calculator (:top-level))
                                     ()
       '(com-arithmetic ,fun))
     (cp:define-command-accelerator ,(intern (format () "COM-~S" fun)) calculator
                                    ,(character fun) () ()
       '(com-arithmetic ,fun))))

;;; The arithmetic command definitions
(define-arithmetic-command +)
(define-arithmetic-command -)
(define-arithmetic-command *)
(define-arithmetic-command /)

;;; The enter command.
(define-calculator-command (com-enter :menu-accelerator "Enter"
                                      :keyboard-accelerator #\Return)
                           ()
  (push current-value value-stack)
  (setq entry-state 'new-no-push))
```

9.5 Fun and Games

From *The Hacker's Dictionary*, Guy L. Steele, Jr., *et al*:

Feature

1. An intended property or behavior (as of a program). Whether it is good is immaterial.

2. A good property or behavior (as of a program). Whether it was intended is immaterial.

3. A surprising property or behavior; in particular, one that is purposely inconsistent because it works better that way. For example, in the EMACS text editor, the "transpose characters" command will exchange the two characters on either side of the cursor on the screen, *except* when the cursor is at the end of a line; in that case, the two characters before the cursor are exchanged. While this behavior is perhaps surprising, and certainly inconsistent, it has been found through extensive experimentation to be what most users want. The inconsistency is therefore a feature and not a BUG.

4. A property or behavior that is gratuitous or unnecessary, though perhaps impressive or cute. For example, one feature of the MACLISP language is the ability to print numbers as Roman numerals. See BELLS AND WHISTLES.

5. A property or behavior that was put in to help someone else but that happens to be in your way. A standard joke is that a bug can be turned into a feature simply by documenting it (then theoretically no one can complain about it because it's in the manual), or even by simply declaring it to be good. "That's not a bug; it's a feature!"

The following list covers the spectrum of terms used to rate programs or portions thereof (except for the first two, which tend to be applied more to hardware or to the SYSTEM, but are included for completeness):

CRASH	BUG	CROCK	WIN
STOPPAGE	LOSS	KLUGE	FEATURE
BRAIN DAMAGE	MISFEATURE	HACK	PERFECTION

The last is never actually attained.

10. Systems, Storage and Errors

This chapter is for people who want to write applications programs and package them professionally. I will be discussing topics which programmers need to understand eventually if they want to write completely self-contained software that is easy to install and run and doesn't scare its users.

10.1 Systems

The *System Construction Tool* (SCT) provides a mechanism for keeping track of multiple files which together make up a single program. You define a system with the **defsystem** macro; a system is made up of files and, potentially, other systems. You can compile a system with the CP command Compile System, and load a system into your Lisp environment with Load System.

Systems can also be distributed on tape to other sites. To do this, you would use the Distribute Systems command. Loading such a tape is performed using the Restore Distribution command.

10.1.1 Defining a System

The **defsystem** macro is used to define a system. Unlike that
for previous releases, the documentation for **defsystem** is both
comprehensive and easy to understand. **defsystem**'s syntax is
as follows:

```
(defsystem name
    (:option1 value1
     :option2 value2 ...)
  module1
  module2
  ...)
```

The *options* are used to declare the overall attributes of the
system. A few of the interesting ones include:

:pretty-name   how to display the name of this system in the
herald.

:bug-reports   to whom to send mail when there is something wrong with this system.

:patchable   whether the system can be changed without
having to recompile the whole thing. See the
section "Patching a System," page 222.

:default-package what package to use if the source files don't
say.

The *modules* describe the files (and systems, if any) which are
part of the system. Strictly speaking, a module defines the
operations which are performed on those files, such as compiling and loading them. To do this, modules have a *type*, which
actually defines the operations. The default type is **:lisp**, although that can be changed with the option
:default-module-type.

In addition, modules define *dependencies, i.e.,* which operations must be performed before an operation may be performed on the files in this module. For example, a file full of macros might need to be loaded before another file which uses those macros can be compiled.

Module descriptions come in two varieties. "Short form" module descriptions specify files and their dependencies in a simple syntax. Most of your module descriptions will be in this form. "Long form" module descriptions are windier, but allow you to specify completely the type of files, the permissible operations, and the exact dependencies, when the short form doesn't do it completely.

Here is a short-form example from the documentation:

```
(defsystem adventure
    (:default-pathname "games: code;"
     :default-package  adventure)
  (:serial "defs" "macros"
          (:parallel "things" "rooms")
          "parser"))
```

This definition says that to compile this system, you first compile the file `games:code;defs.lisp`, then load it; then compile and load `macros`, then compile and load the files `things` and `rooms` in unspecified order, and then compile the file `parser`.

You will need to use a long form module description to do any of the following:

1. Load one or more files into a different package from the rest of the system.

2. Specify a different module type from the rest of the system (the default module type is **:lisp**; others are shown in

the documentation.   See the section **"defsystem
Operations"** in *Program Development Utilities*.)

3. Specify a more complicated dependency relationship than
   is possible using **:serial** and **:parallel**.

4. Specify that a file need only be loaded if you need it for
   some other operation; *e.g.*, if you only need macros for
   compiling your other files, but not for loading them.

Here is a moderately complicated example from the IP-TCP system declaration file:

```
(defsystem ip-tcp
    (:maintaining-sites :scrc
     :pretty-name "IP-TCP"
     :default-pathname "SYS: IP-TCP;"
     :advertised-in (:herald :disk-label)
     :patchable t
     :distribute-sources t
     :distribute-binaries t
     :source-category :basic)
  (:module components (tcp tcp-service-paths
                           ip-tcp-applications
                           ip-tcp-doc)
           (:type :system))
  (:module notice-text "sys:site;notice.text"
           (:type :text))
  (:serial "chaos-unc-interface"
           "ip-global"
           (:parallel "ip" "ip-routing")
           (:parallel "icmp" "udp" "egp")
           components
           notice-text))
```

```
(defsubsystem tcp
    (:default-pathname "SYS: IP-TCP;"
     :pretty-name "TCP"
     :distribute-sources t
     :distribute-binaries t
     :source-category :basic)
  (:module text-files "tcp-structure"
           (:type :text))
  (:serial (:parallel text-files "tcp-defs")
           "tcp-error"
           "tcp"
           "tcp-user"
           "tcp-debug"
           "distribution")))
```

[There are three more subsystems declared in this file, namely tcp-service-paths, ip-tcp-applications and ip-tcp-doc.]

10.1.2 Compiling and Loading Systems

To compile your system, write your **defsystem** form into a file, and use the CP command Compile System. Similarly, to load your system, use the command Load System. Other operations are currently only available with function interfaces: **sct:hardcopy-system**, for example.

Of course, I have begged one important question: how does the Lisp Machine *find* the file in which you've put your **defsystem** form? This is a job for a file in the site directory, SYS:SITE; *system-name*.SYSTEM. This file should contain a **sct:set-system-source-file** form. Since it's good practice to define systems using logical pathnames, that file might also want to include a **fs:make-logical-pathname-host** form.

For example, here is the file which describes where to find the

"Grapher" example system:[1]
sys:site;grapher.system

```
;;; -*- Mode: LISP; Syntax: Common-Lisp; Base: 10 -*-

(fs:make-logical-pathname-host "LISP-LORE")

(sct:set-system-source-file
  "GRAPHER"
  "LISP-LORE:EXAMPLES;GRAPHER;GRAPHER-SYSTEM")
```

It causes the logical host LISP-LORE to be defined the first time
it is loaded. Then it informs SCT where to find the **defsystem**
form for that system.

10.1.3 Patching a System

When a system has been declared patchable, that means that
you can make incremental changes to it. These changes, called
patches, are loaded after the system itself, and make modifications to the Lisp world, thus redefining whatever functions, *etc.*,
have been patched.

To create a patch, use the editor command m-X Start Patch.
You will be asked for the name of the system you wish to
patch. After your patch has been started, you may add as
many definitions to the patch from your editor buffers as you
wish. When you're done, the command m-X Finish Patch writes
out the patch file, compiles it, and updates the *patch directory*.
After you're done patching your system, you should remember
to save the file buffers you change; otherwise, next time you
recompile the system you will lose the changes you've made.

---

[1]The rest of the system is in Chapter 7. Logical pathnames were discussed in
chapter 8. See the section "Logical Pathnames," page 195.

How do you add patches? Well, first you should modify the definitions in your editor buffer. Compile them with c-sh-C and test them out, to make sure they work as you expect. Next, use the m-X Add Patch command. This command adds the definition at the cursor to the patch file. As it does so, it asks you for a comment about the change you are making.

You can also add patches wholesale. You can use m-X Add Patch with a region set to add the contents of the region to your patch. Also, the command m-X Add Patch Changed Definitions and its relatives will scan one or more buffers to see what you've changed, and offer to patch those definitions, one at a time.

When you're done, use the command m-X Finish Patch. Your patch will be written out and compiled, and the system patch directory will be updated to remember the patch comments.

Sometimes you decide afterward that a patch was not correct after all. You can get rid of it before you finish it with m-X Abort Patch. If you've already finished it, you can still abort it, but first you have to pretend you never finished it: use m-X Resume Patch. You can also correct mistakes in patches with m-X Edit Patch File.[2] After you save the file, you should compile it with m-X Recompile Patch rather than m-X Compile File or the CP equivalents, because m-X Recompile Patch will compile it with the right logical pathname inside the file.

If a system has been patched since you loaded it, you can add its patches to your world. Use the Load Patches CP command. If a patch has been recompiled, you might like to use m-X Reload Patch in the editor.

---

[2]Of course, you can always just patch the same definition again with the correct version.

10.2 Storage Allocation

Every Lisp program creates and discards objects. The nature of Lisp is such that programmers need not care about storage allocation, as when you let go of an object, it just "disappears."

Well, so much for a nice fantasy. While all of this is true, storage allocation can be one of the nastiest sources of inefficient program execution. Let's discuss storage allocation more fully.

10.2.1 Allocation and the Garbage Collector

Your programs don't *have* to do anything interesting to use storage allocation. Pretty much everything you do will create new Lisp objects. Every time you create a new object, it gets put in the default place. The *garbage collector* (GC) will come along eventually and reclaim the objects you are no longer using. This is the ideal Lisp allocation scheme. After all, why burden the programmer with details she left behind with Fortran?

There are two inefficiencies with using this scheme. First of all, allocating objects takes time, and might cause page faults.[3] Secondly, the garbage collector itself takes time, and also might cause page faults.

The other problem with most garbage collectors are their unpredictability. The garbage collector can start up at any time

---

[3] Page faults are what happens when you refer to a part of your environment which is not actually in the machine's normal, physical memory. The machine comes to a screeching halt, orders up the *page* of *virtual memory* containing that object from the disk, and waits for the disk to give it the data. This can take a long time, on the order of milliseconds. It's not good for performance to take many page faults.

the allocator decides that storage is getting tight, and can take an arbitrarily large amount of your computing resources to do its job. While a great deal of thought has gone into the controls for the GC, once in a while it will start up right when you need as much Lisp Machine as you can get.

In recent releases, Symbolics has added an optimization of the standard garbage collector, called the *ephemeral GC*, or EGC. Its job is to clean up potential garbage in recently allocated storage, and be as quick and unobtrusive about it as possible. It uses very little of your machine resources, and can therefore be run frequently, without your noticing it. In fact, for many applications it actually *improves* the overall performance of your program, because it tends to bring related structures closer together in memory, reducing the number of page faults you take.

The full-blown, *dynamic* GC is responsible for the rest of the garbage, and gets run as infrequently as possible. Many people turn the EGC on and the Dynamic GC off. I usually run my Lisp Machine that way. The EGC, while just as unpredictable as the Dynamic GC, is nearly always over before you notice it. Most applications only need the Dynamic GC if they are consing up[4] a lot of objects. A compromise between always running the Dynamic GC and running with only the EGC is to cause a Dynamic GC at some time when you don't care about how much machine it uses up. The function **gc-immediately** will do just that; you can either run it just before you go home at night (or whenever you *do* go home), or you can have a background process which wakes up once a night and decides if you're not using your machine, and then runs the GC.

---

[4]This hacker's term is defined in the next chapter. See the section "Fun and Games," page 277.

10.2.2 Areas

One way to decrease the number of page faults your machine takes is to increase the number of useful objects you get with each one. In general, virtual memory systems behave better when you increase the *locality* of your references between objects. Thus, it is better to have related objects allocated together, if possible.

Normally, you don't take control over where your objects get allocated. There is a way to do so, however. What you do is to tell the system what *area* to use when it allocates your objects. An area is a storage allocation space. While objects in the same area are not guaranteed to reside on the same page, the likelihood is much greater than if you don't take any control over allocation at all. Also, the paging system is more efficient for objects residing on *adjacent* pages, since a page fault will often cause several pages on one or both sides of the missing page to get read in at the same time, which is faster than reading them in one at a time.

To create an area, use the function **make-area**. It takes a number of keywords; the only required one is **:name**, which must be a symbol. The value of that symbol is set to a handle to the area, called its *area number*. You will probably only want to create an area once, perhaps using **defvar** or by making it be a **:once** initialization.[5]

Most of the storage allocating functions either permit you to specify the area directly or have variants which take an area as an argument. For example, **make-instance** and **make-array** have **:area** keyword arguments. **cons** and **list** have twin functions named **cons-in-area** and **list-in-area**.

---

[5]See the section "Introduction to Initializations" in *Internals, Processes, and Storage Management*.

For example:

```
(defvar *my-area* (make-area :name '*my-area*))

(defun create-a-foo (&rest options)
  (apply #'make-instance 'foo :area *my-area* options))
```

There is also a global variable named ***default-cons-area***, which is used if you don't specify an area in which to allocate storage. You can bind that variable to your area, but be careful: other "system" allocations can go on behind your back, and get stuck into your area.

10.2.3 Resources

Resources are actually quite well documented in volume 8, so I will just discuss the highlights.

In cases where a program creates and then discards large objects at a high rate, it can be worthwhile to do the storage management manually, rather than relying on the garbage collector eventually to clean up. The *resource* facility provides a simple way to do so, and is widely used throughout the system software. The window system, for example, allocated and frees certain kinds of windows (which are very large objects) moderately often. It uses resources for this.

For each resource defined, there is conceptually a list of free objects "in" that resource. *Allocating* an object from a resource involves checking the list of free objects and returning one if any is suitable; if not, a new one is created and returned. *Deallocating* an object involves placing the previously allocated object in the free list. The storage space occupied by a deallocated object is not really freed in the sense that the GC can claim the space; it does not become available to be used as part of a newly created Lisp object. The original object continues to occupy the storage space, but may itself be reused by being allocated again.

The four functions and macros which compose the programmer interface to the resource facility are:

- **defresource**, for defining new resources;
- **allocate-resource**, for allocating an object from a resource;
- **deallocate-resource**, for freeing an allocated object; and
- **using-resource**, which temporarily allocates an object and then deallocates it.

A call to **defresource** looks like this:

```
(defresource name parameters
  keyword value
  keyword value
  ...)
```

name should be a symbol, which will be the name of the resource. *parameters* is a (possibly empty) list of pseudo-arguments which will be used to determine which free objects are actually suitable, and to allow the constructor to create a good one if none are free. For example, a resource of two-dimensional arrays might have two parameters, the number of rows and the number of columns. When allocating an object from this resource, you could specify how many rows and columns it should have. The free list would be filtered for arrays with the requested dimensions – if all arrays on the free list had the wrong dimensions, a new one would be created.

There are seven possible keyword options. Only one, the **:constructor** option, is required.

- **:constructor** – defines the function which creates a new object.

- **:initializer** – a function which is called when an object is allocated, whether newly created or reused. If not specified, it defaults to a no-op.

- **:checker** – a function which determines whether it is safe to allocate an object. All objects in a resource, whether the resource facility has been told they have been deallocated or not, are considered by the checker; the default checker tests for whether the object has been deallocated.

- **:matcher** – a function which determines whether a free object is suitable, according to the *parameters*.

- **:finder** – a function which bypasses the entire allocation scheme: the checker, matcher and constructor are all folded into this one function.

- **:initial-copies** – a number which tells the resource facility how many objects to create in advance.

- **:free-list-size** – a number which informs the resource facility how many objects you expect to create. The "free list" is actually kept as an array which contains both allocated and deallocated objects, and this number is the default array size. **:free-list-size** is actually a bit of a misnomer.

The rest of the resource facility is concerned with allocating and deallocating objects:

- **allocate-resource** *resource-name* &rest *parameters*
  An object is allocated from the specificed resource, matching the given parameters. A second value, a resource descriptor, is returned, but may be safely ignored.

- **deallocate-resource** *resource-name* *object* &optional

descriptor
The object is returned to the resource's free-list. If you saved the resource descriptor from **allocate-resource**, passing it back now will make deallocating much faster.

- **using-resource** (*variable resource parameters* ...) *body* ...
This macro, which calls *allocate-resource* and *deallocate-resource*, is preferred to calling those two functions directly. The *body* forms are evaluated inside a context where *variable* is bound to an object allocated from *resource* with the specified *parameters*. The object is deallocated at the end. An **unwind-protect** is used to guarantee that the object is deallocated; **using-resource** returns the value of the last form in the *body*.

Now an example. We define a resource of raster arrays, with parameters for the number of rows and columns, which default to 128 each. A matcher is provided which accepts any array whose dimensions are at least as great as the given parameters (the default matcher would require that the dimensions be exactly the same, meaning that we would very rarely reuse an object.) And an initializer fills the array with zeros.

```
(defresource sloppy-raster
             (&optional (rows 128) (columns 128))
  :constructor (make-raster-array rows columns
                                  :element-type
                                  '(unsigned-byte 1))
  :matcher (multiple-value-bind (width height)
               (decode-raster-array object)
             (and (≥ columns width) (≥ rows height)))
  :initializer (bit-xor object object object))
```

And, to use our resource:

```
(defun mangle-window (window)
  (multiple-value-bind (width height)
      (send window :inside-size)
    (using-resource (sheet-array sloppy-raster width height)
      (send window :bitblt-from-sheet tv:alu-ior
            width height 0 0 sheet-array 0 0)
      (mangle-window-contents sheet-array)
      (send window :bitblt tv:alu-ior
            width height sheet-array 0 0 0 0)))))
```

Debugging functions for use with resources include **deallocate-whole-resource, clear-resource** and **map-resource**. I encourage you to read the documentation for these functions if you are interested.

10.2.4 Stack Allocation

There is one other place you can use to allocate objects you know will be very short-lived. The stack can be used to hold these objects.[6] There is only one problem with this: you have to be careful what you do with these objects you create, because they *go away* when you return from the current function invocation.

Let's start with an example. Suppose you have a variable named ***search-space*** which is ordinarily a list of all the rooms you want to search for your algorithm. You might want to restrict your search temporarily to, say, a list of three objects:

---

[6]There are actually *three* stacks which are managed by the Lisp Machine in parallel: the *control* stack, which is used for function calling, arguments, and temporary variables; the *binding* stack, which is used for special variable bindings, and the *data* stack, which is only used for stack-allocated objects. You don't need to understand this for the discussion which follows.

```
(defun search-apartment (object)
  (with-search-restricted (livingroom bedroom kitchen)
    (search-for object)))
```

Now, one way to do this would be to write your macro **with-search-restricted** in the normal way:

```
(defmacro with-search-restricted ((&rest rooms) &body body)
  '(let ((*search-space* (list ,@rooms)))
     ,@body))
```

However, if you're going to be doing this a lot, you don't want to create those lists every time you use this macro, only to be dropped on the floor. Instead, you can allocate the list on the stack:

```
(defmacro with-search-restricted ((&rest rooms) &body body)
  '(stack-let ((*search-space* (list ,@rooms)))
     ,@body))
```

In general, the macro **stack-let** knows how to deal with all kinds of lists and arrays. In a future release, it is supposed to know about instances as well. However, if **stack-let** doesn't know how to create the kind of object you want on the stack, it will turn into a normal **let**.

Remember that warning about objects going away. It's very important that you not *store* any of these objects in permanent storage, or return them as values from your functions. For example, don't do this:

```
(defun tear-up-rooms (livingroom bedroom kitchen)
  (stack-let ((apartment (list livingroom bedroom kitchen)))
    (tear-up apartment)
    apartment))
```

(it returns a stack object) or

```
(defun create-apartment (livingroom bedroom kitchen)
  (stack-let ((apartment
                (make-array 3
                  :initial-contents
                  '(,livingroom ,bedroom ,kitchen))))
    (setq *the-apartment* apartment)))
```

(it stores a stack object in permanent storage) because your Lisp Machine might crash or make the garbage collector very confused.

10.3 Condition Handling

The *condition* system is also very well documented (See the section "Conditions" in *Symbolics Common Lisp: Language Concepts.*) Again, this section will just be an overview to get you started.

The whole idea behind the condition system is that you should be able to write your programs assuming that everything will work properly, and then later provide other code which will be used when something unusual happens. A condition is *signalled* when such an *event* occurs.

A *handler* is a piece of software which is invoked when a condition is signalled. Handlers can examine the condition and determine the response to it: enter the debugger, retry using a different argument, discard the attempt to do whatever caused the error, and so forth. Handlers can be either global or dynamic in scope. The system provides default global handlers for all conditions: errors, for example, always enter the debugger unless a dynamic handler overrides that response.

Certain conditions are errors, such as an attempt to divide by zero, or to open a file which doesn't exist. Other conditions are not errors, but a program might like to know that they have happened anyway, so as to make the machine do something other than what it usually does. These conditions include an attempt to open a file when you're not yet logged into your Lisp Machine, or when your Lisp Machine is about to ask you a question. Perhaps your program has anticipated that you might be asked a question, and "knows" the answer: it might provide a handler which examines the question to make sure it's the right one, and then supplies the answer.

The condition mechanism is built on flavors, of course. There are three ways to customize the condition mechanism for your programs:

1. Signalling existing flavors of conditions within your code, which may invoke the system's default handlers or ones that you've written.

2. Defining handlers for existing flavors of conditions which may be signalled by system code.

3. Defining new flavors of conditions, which you may then signal, and for which you may then write handlers.

10.3.1 Signalling Conditions

The mechanism for signalling conditions relies on flavors. Each class of events corresponds to a flavor which is built on the flavor **condition**. Signalling a condition involves creating an object of the appropriate flavor and then running through the appropriate handlers, finding one that will accept the condition.

For example, when you attempt to divide by zero, the condition object created is an instance of the flavor **sys:divide-by-zero**. The instance variables of the condition object will contain information that describe the event.

Every condition object has certain generic functions defined on it. For example, **dbg:report-string** returns a string which the debugger prints upon entry. **dbg:report** takes a stream and prints that string on the stream (you can also use **princ** or format with ~A for this effect).

You can signal a condition with **error** or **signal**. If you want to allow your handlers to do interesting things, you might try investigating **signal-proceed-case**. See the section "A Few Examples," page 240.

10.3.2 Handling Conditions

Each handler is defined to be applicable only for one flavor of condition object. It is invoked by the signaller when a condition of that flavor, or one built on it. The set of conditions a handler can handle is thus determined by the flavor inheritance mechanism. Dynamic handlers (*i.e.*, not global ones) have dynamic scope, so finding a handler for a given condition involved stepping back through the stack and invoking handlers which are applicable to that condition until one of them actually handles it. If no dynamic handler will accept the condition, the signaller uses the global handlers instead.

There are several kinds of actions a handler can take.

- It may decline to handle the condition at all; in this case, the signaller continues searching the stack or global handlers for another handler.

- It may instruct the program to continue past the point where the condition was signalled, possibly after correct-

ing the circumstances that led to the event being signalled. This is called *proceeding*.

- It may unwind the stack to the point where the handler
  was bound, flushing the pending operations. This behavior is essentially equivalent to what you'd get with a
  catch in place of the handler, and a **throw** – with the
  correct tag – in place of the signalling of the condition.

- It may partially unwind the stack to some intermediate
  point and re-execute from there. This kind of handler is
  called a *restart* handler.

You can write condition handlers with one of the following special operators: **condition-case, condition-bind, condition-bind-
default** or **condition-call.**[7] A special kind of handler is provided
with **ignore-errors**, which merely returns **nil** if any error is
signalled within its body. See the section "Bound Handlers" in
Symbolics Common Lisp: Language Concepts.

10.3.3 Creating New Condition Flavors

Condition objects are instances of the flavor **condition.** Two
more specialized flavors are built on top of **condition:**

- **error** – an error condition.

- **dbg:debugger-condition** – causes debugger to be entered
  when signalled. This flavor is only here so you can write
  handlers for **error** which don't trap certain error conditions, such as those that the debugger uses internally.
  error is built on **dbg:debugger-condition.**

---

[7]Each of these has a conditional form, *e.g.*, **condition-bind-if**, which only creates
a handler if a predicate you provide tests true.

You should probably make all your conditions be based on the flavor **error**.

All condition flavors must define a method for the generic function **dbg:report**.[8] It should take one argument, a stream, and print its message on that stream.

Some program examples will appear later in this chapter. See the section "A Few Examples," page 240.

10.3.4 Restart Handlers

When a computation blows up into the debugger, a programmer might want to allow the user to restart the process at a controlled point, rather than just wherever it happened to die. This is what a *restart handler* is for. A restart handler shows up as one of the options in the debugger.

For example, suppose you're copying a file via a network connection and the remote host goes down. An error built on **sys:network-error** would be signalled. Rather than trying to pick up where the error is signalled, in general you would need to open the file again from the beginning and start over. Functions such as **copy-file** contain such restart handlers.

The basic restart handler is created with **catch-error-restart**. You supply a set of conditions to which it applies. If the debugger is entered and the user selects the corresponding option, **catch-error-restart** returns **nil** as its first value, and a non-**nil** value as a second value. Otherwise, it returns the value of its last form.

Fancier restart handlers can be written using **error-restart** and

---

[8]In old code, you might see a programmer define a method for the message **:report**, which was the old way to do this.

error-restart-loop. I recommend you look at the documentation for these. See the section "Restart Functions" in *Symbolics Common Lisp: Language Concepts.*

[For advanced users: if you want to invoke restart handlers from a handler established with **condition-bind,** you have two choices: you can take pot luck and get whatever handler the function **dbg:invoke-restart-handlers** chooses, or you can be fancier. The restart handlers are in a global variable named **dbg:*restart-handlers*.** They accept the following messages:

- **:describe-restart** *stream* – prints the string that appears in the debugger for that restart handler.

- **:handle-condition-p** *condition* – tells you whether the restart handler is supposed to handle the given condition.

- **:handle-condition** *condition tag* – invokes the restart handler. How you get the tag is complicated. See the source to **dbg:invoke-restart-handlers.**]

10.3.5 Proceeding

A cogent five-page discussion of what is involved in programming proceedable errors is to be found in the documentation: See the section "Proceeding" in *Symbolics Common Lisp: Language Concepts.* I recommend reading it. I will include just a few highlights here.

For proceeding to work, two conceptual agents must agree on a protocol:

- The program that signals the error.

- The **condition-bind** handler that decided to proceed from the condition, or else the user who told the debugger to proceed.

The signaller signals the condition and provides a set of alternative *proceed types.* The handler choose from among the proceed types to make execution proceed.

A proceed type is *defined* by giving the condition flavor a **sys:proceed** method. Since **sys:proceed** methods are combined with :**case** method combination, a condition flavor can have any number of **sys:proceed** methods, each defining a different proceed type. The first argument to the generic function **sys:proceed** (after the condition object, of course) is a dispatch argument which selects the actual method.

The body of the **sys:proceed** method can do anything it wants, generally trying to repair the state of things so that execution can proceed past the point at which the condition was signalled. It may have side-effects on the environment, and may return values (which will be returned by **signal**), so that the function that called **signal** can try to fix things up. Its operation is invisible to the handler; the signaller is free to divide the work between the function that calls **signal** and the **sys:proceed** method as it sees fit.

An easy way to signal proceedable errors is with the macro **signal-proceed-case.** This signals the error, providing only those proceed types which it is willing to handle, and then dispatches on the returned value.

10.3.6 A Few Examples

```
(condition-case ()
    (/ a b)
  (sys:divide-by-zero *infinity*))
```

This form binds a handler for the **sys:divide-by-zero** condition, and evaluates (/ a b) in that context. If the division finishes normally, its value is returned from the condition-case. If **b** turns out to be zero, the **sys:divide-by-zero** condition is signalled, and out handler is invoked, which simply causes the **condition-case** to return the value of the symbol ***infinity***.

```
(condition-case (result)
    (do-something-interesting)
  (error (format *error-output*
                 "Something intersting failed: ~→~A~←"
                 result))
  (:no-error (do-something-else-to result)))
```

In this case, we call **do-something-interesting** with a condition handler for *all* errors. If it works, we can do something else to the returned value. Otherwise, we print the error message on ***error-output***.

```
(condition-bind ((sys:divide-by-zero
                   (lambda (condition)
                     (sys:proceed condition
                                  :return-values
                                  (list *infinity*)))))
    (/ a b))
```

has the same result as the first case above. The proceed type **:return-values** for the condition **sys:divide-by-zero** takes a list of values to return from the division instruction.

```
(defun copy-file-until-you-get-it-right (from-path to-path)
  (loop until (ignore-errors
                (copy-file from-path to-path)
                t)))
```

This is a simple program to retry copying a file until it succeeds, which is a handy thing to have when your network is
flaky:[9]

```
(defun simple-copy-file (from-path to-path)
  (error-restart ((sys:network-error)
                  "Retry copying ~A to ~A" from-path to-path)
    (with-open-file (from from-path)
      (with-open-file
        (to to-path
            :direction :output
            :element-type (send from :element-type))
        (stream-copy-until-eof from to)))))
```

This establishes a restart handler which retries to copy the file
if you have a network error. You would have to press Resume in
the debugger to invoke this restart handler.

Here is a sample program which uses proceeding. Some notes:

- Even though there are two proceed types for this error
  flavor, only those which are explicitly listed in the
  signal-proceed-case are offered to the user. If you don't
  want to allow all of the possible proceed types for a given
  error, you don't have to do so.

- When your proceed case wants to take values, you should
  make them optional and allow your function to prompt for
  them, if you ever want to use it from the debugger. The
  debugger will not pass any values to your proceed method.

---

[9]See hacker's definition at the end of the chapter.

```
(defvar *block-colors* '(red blue green))

(defflavor block-wrong-color
        (block color)
        (error)
  :readable-instance-variables
  :initable-instance-variables)

(defmethod (dbg:report block-wrong-color) (stream)
  (format stream "The block ~S was ~S, which was ~
                  an invalid color."
          block color))

(defmethod (sys:proceed block-wrong-color :add-new-color) ()
  (push color *block-colors*)
  (values :add-new-color))

(defmethod (dbg:document-proceed-type block-wrong-color
                                        :add-new-color)
          (stream)
  (format stream "Add ~A to the list of valid colors" color))

(defmethod (sys:proceed block-wrong-color :ask-for-new-color)
          (&optional (new-color
                        (accept '((member ,@*block-colors*))
                                :prompt "New Color")))
  "Supply replacement color"
  (values :ask-for-new-color new-color))

(defflavor block (location color)
          ()
  :initable-instance-variables)
```

```
(defmethod (make-instance block) (&rest ignore)
  (unless (member color *block-colors*)
    (signal-proceed-case
      ((new-color)
        'block-wrong-color :color color :block self)
      (:add-new-color)
      (:ask-for-new-color (setf color new-color)))))

(defun make-block-force (color location)
  (condition-bind ((block-wrong-color
                    (lambda (condition)
                      (sys:proceed condition
                                   :add-new-color))))
    (make-instance 'block :color color :location location)))

(defun make-block-valid-color (color location)
  (condition-bind ((block-wrong-color
                    (lambda (condition)
                      (sys:proceed
                        condition
                        :ask-for-new-color
                        (first *block-colors*)))))
    (make-instance 'block :color color :location location)))
```

10.4 Fun and Games

Flaky, Flakey *adjective.*
    Subject to intermittent failure.
    This use is, of course, related to the common slang use of
    the word, to describe a person as eccentric or crazy. A
    system that is flaky is working, sort of, enough that you
    are tempted to try to use it; but it fails frequently enough
    that the odds in favor of finishing what you start are low.
    From *The Hacker's Dictionary*, by Guy L. Steele, *et. al.*

11. The Card Game Example

This program grew out of a discussion I had with Muffy Barkocy after seeing her Solitaire program for earlier releases. The basic idea of this program is to provide a solitaire-program substrate. The idea was to enable programmer to implement a new solitaire game in an hour or so.

As before, if your site has loaded the tape that comes with the book, do Load System Cards to load the code. Type Select Square to get to the game itself.

There are four more-or-less independent parts of this system. These are:

1. the definitions of cards and card-playing behavior in general;

2. the definition of "places" in which cards are to be played (and displayed);

3. the redisplay; and

4. the definitions of the various games which have been implemented.

I will consider each of these in the sections which follow.

By the way, the card drawings were consed up[1] using the Symbolics Font Editor by Muffy Barkocy, with minor additions by myself. They are stored in the font **fonts:deck-of-cards**. The program display looks like figure 7.

11.1 Card Definitions

The basic parts of the card game database are **cards** and **card-decks**. A **card** is implemented as a flavor instance, and has a rank (a number between 1 and 13) and a suit (a keyword symbol). Its other instance variables are all related to drawing the card on the screen. The **glyph** is a number which is the index into the font **fonts:deck-of-cards** of the glyph used to draw the card. The **color, draw-alu** and **overlay-alu** instance variables are used to draw the cards in color (the color implementation will not be discussed in this chapter).

The two functions of interest in drawing cards are **display-card** and **erase-card-spot**. The first draws a picture of the card in its appropriate position (and color(s), as necessary). The second clears the background away for the purpose of drawing the card. Note that it draws a "black" glyph (the solid card back) using either the "erase" alu (for B/W screens) or the "white" alu (for color screens).

One other drawing primitive is also in the file. **display-empty-place** is used for displaying other glyphs from the card deck font. This is how card backs, white or black spaces, and jokers are drawn, since they are not cards *per se*.

Card decks are arrays with structures in their leaders. This is

---

[1] See hacker's definition at the end of chapter.

Figure 7.   Card game program display window

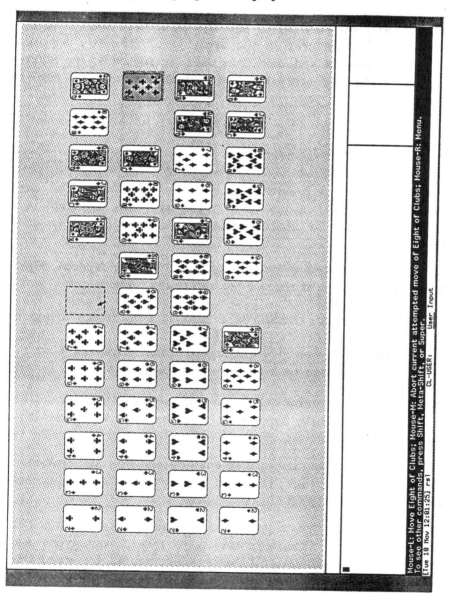

a common technique for adding structured data to arrays. Making a structure of type **:array-leader** (or **:named-array-leader**) gives you the structured data, and still allows you to use **aref** on the array part. The two structure elements, **n-cards** and **next**, are used, respectively, for allocating the right size array and for dealing out the next card in the deck.

11.2 Presentation Types

There are several presentation types in this file. The **card** presentation type was not used in this program, although it is available for use by programmers who wish to use it to write new games.

Its parser is the most interesting part. It uses **accept** recursively, reading one of three types internally. These are:

- A rank, the string "of", and a suit (*e.g.*, "King of Spades").
- A rank followed by a suit (*e.g.*, "Three Diamonds").
- A suit followed by a rank (*e.g.*, "Heart four").

Note that the recursive call to **accept** provides the keyword argument **:prompt nil**. Try seeing what happens when you omit this argument.

The two internal routines **parse-card-description** and **find-card** are used to figure out the rank and suit from the input, and look the card up (or create it), respectively.

The describer is used to prevent the default describer from getting into the act. The describer for **or** presentation types goes through, in loving detail, all the possible inputs. I decided that this was undesirable.[2]

The remaining presentation types are used for presenting card places on the screen. There are two types: **card-places** are places where cards have actually been displayed. **empty-places** are places where cards might be played.

11.3 Card Places

A card place is a data abstraction used to signify a place where a card has been played, or where one might be played.

11.3.1 Basic Places

Each place has a (potentially empty) list of cards in it, called its contents. Every place also has a number of rules associated with it:

- *addition rules* determine whether it is legal to add a certain list of cards to that place.

- *addition side effects* cause things to happen when cards are added to a place.

- *removal rules* determine whether it is legal to remove the contents of a given place.

- *removal side effects* cause things to happen when cards are removed from a place.

The **basic-card-place** flavor implements all of these. The three

---

[2]Try commenting it out and you'll see why.

methods which implement these are **contents-may-be-removed,**
contents-may-be-stored, and the **:after** method of
(setf place-contents). Note that you don't want to perform
the side effects when you're creating the card place object, so
the special variable ***no-side-effects*** is used to prevent that
from happening.

11.3.2 Presentation

Presenting the contents of a card place is the job of
self-presenting-mixin. This mixin presents two parts of the
place independently, namely the cards (the contents) and the
empty places. The reason it has to be so circuitous is that
some places will contain both cards and empty places, such as
stack-places, which contain some cards, and finally an empty
place at the bottom into which you might play one or more
other cards.

This mixin remembers the presentations for the cards and
empty places separately. One reason to do this is for the incremental redisplay. The redisplay needs to remember where
every display is, so as to re-draw it as necessary. It also needs
to remember the presentations so as to erase those presentations from the screen. It also uses an internal method of
presentations **dw::presentation-mouse-sensitive-boxes** to find
out exactly where the presentation is on the screen, so as to
"erase" them by overdrawing them with a background rectangle. Highlighting and unhighlighting of card places is also
done using the information about the presentation itself.

Finally, this mixin is responsible for making sure that all
places whose contents have changed get redisplayed. It calls
the routine **redisplay** on itself and all its superiors, which
schedules all those places for redisplay. It has a daemon on
the **(setf place-contents)** method for that purpose.

11.3.3 Caching

After I ran the game for a while, I discovered it was creating and discarding lists and values all over the place. The mouse-sensitivity testing was also invoking the addition and removal rules methods quite frequently, and most of the time the result of that calculation was being dropped on the floor.

Thus, I created two mixins, **contents-remembering-mixin** and **rules-remembering-mixin**, which store the contents and results of running the rules methods, respectively. These mixins do not change the results of those methods, but just save the values (and the conditions under which they are valid, in the case of the addition rules) away for reuse.

11.3.4 Stacked Places

The normal card place, implemented with the flavor **card-place**, can either be empty or hold exactly one card. The **place-contents** method and its **setf** equivalent store and obtain the contents in the **card** instance variable (the **empty-appearance** I.V. is what the place looks like when empty, and can be any of the symbols in ***empty-place-glyph-info*** or **:invisible**). In addition, the **contents-may-be-stored** method enforces the requirement that only a single card may be stored in a one-card place, and then only if it's empty. **card-place** also has methods for **present-contents** and **present-empty-spaces**, which hook it up to **self-presenting-mixin**.

However, in many solitaire games, you actually want to have more than one card in a place, and have them behave as a stack which must be moved together or not at all. This is implemented using a **stack-place** instance. A stack place only remembers two things, just like a cons: the head of the list and the rest. When you ask for the contents of a stack place, it returns a list whose head is the head card, and whose **cdr** is the contents of the rest of the stack. Setting the contents of a **stack-place** involves making new instances for the places which represent the rest of the stack, much akin to creating new cons cells when you copy a list.

One peculiarity of stack places is the way they present themselves. It's done recursively: first you present the card at the top of the stack, and then present the rest of the stack. With the whopper on **present-self** in the mixin **self-presenting-mixin**, this causes the overlapping presentations to become part of the overall presentation. Thus, when you point the mouse at the (presumably partially overlapped) card at the top of the stack, the whole stack becomes mouse-sensitive (assuming it's legal to remove the contents of the place).

A minor variant of a **stack-place** is a **stack-with-face-down-cards-place**. In addition to all the other attributes of a stack place, a stack place with face down cards has a special removal side-effect: if there are any face-down cards, one of them becomes the new contents of the place. The Spider game uses this kind of place for each of its piles.

11.4 The Interactive Program

The interactive part of the program is defined using the macro **dw:define-program-framework**, which is again written using the layout designer "Frame-Up" (Select Q). The command table claims not to inherit commands from any other command table; actually, the **:inherit-from** list is supplied at run time, and depends on the game you're playing.

Once again, we have turned off command echoing. The echo is very distracting.

There are two command menus in the window layout. One of them, the menu for the :games menu level, will contain commands which start different kinds of games. The other is used for normal commands.

11.4.1 Games

The state variable of a card table program is the current **game**. Since this variable is used in a number of different contexts, I have defined three macros which can always be used to get at it. They are both called **current-game**. One is a vanilla defmacro, and is used in most code outside of the system. The others are defined with **defmacro-in-flavor**, and make the game accessible from inside methods of the program itself (e.g., commands) and from methods of the game (where, unsurprisingly enough, the macro expands into a reference to **self!**).

Each game instance is a flavor defined on the low-level flavor **basic-game**. The basic game flavor remembers several interesting bits of state:

- A list of card places which are to be displayed for the game.

- A list of card place descriptions for the game. When you create a game, this list is filled in using the **make-card-place-descriptions** method of the game. This list, in turn, is used to create the list of card places, and also contains information about where they are to be displayed.

- A pair of card decks, one "raw" and one shuffled. The latter is made by calling **shuffle-deck** on the former at the beginning of each game.

- A "place to move:" when you click on a card place whose

contents you want to move, but have not yet moved it by clicking on the empty place you want to move them to, the "from" place is remembered here. The macros **place-to-move-valid** and **place-to-move** permit access to this variable from whatever code you happen to be running.

Each time you start a new game, you re-make all its card places from the card place descriptions. The method **make-card-place** is defined using the :case method combination type. The three common ones are defined on **basic-game**: :card, :stack and :stack-with-face-down-cards.

A few commands are defined for all games. The commands which allow you to specify the "from" and "to" places for a move, for example, are defined here. The "Oops" command, meaning that you didn't really mean you wanted to move that card, is also defined. The "new game" command is here, and also the "next round" command. Finally, the command which allows you to move the game to the color screen is also defined here.

Finally, the macro **define-game** is defined here. This macro causes the following definitions to be done:

- A command table for the command is created.

- A game flavor is defined.

- A command definer for the game is created.

- A command which selects the game is defined.

All game definitions begin with an invocation of **define-game**.

11.4.2 Place Display

The game board pane is defined to have a redisplay method named **display-game-board**, which is invoked after each command. It explictly says that this function is an incremental redisplay function, but doesn't use the system-supplied redisplay functions (**:incremental-redisplay :own-redisplayer** is how you say this). I didn't use the **:incremental-redisplay t** mechanism because the normal redisplay technology doesn't work for overlapping displays.

The variable ***redisplay-list*** contains a list of places which must be redisplayed, because their contents have changed. Items are placed on this list by the fuction **redisplay**. The primary caller of this function is the **setf** method for self-presenting places. Any function which wants to force a complete redisplay calls the function **complete-redisplay**, which schedules a complete redisplay.

The display function for the game board determines whether a complete redisplay is to be done; if so, it erases the window (using the **:clear-history** message), draws the background rectangle, and tells the game to display itself. If an incremental redisplay can be done, only those places on the redisplay list are told to do so.

11.5 The Program

lisp-lore:examples:card-game:card-system.lisp

```
;;; -*- Mode: LISP; Syntax: Common-lisp; Package: USER; Base: 10 -*-

(defpackage cards
  (:use SCL)
  (:colon-mode :external))

(defsystem cards
    (:patchable t
     :default-pathname "lisp-lore:examples;card-game;"
     :maintaining-sites :ssf
     :pretty-name "Card demo program")
  (:module font ("deck-of-cards") (:type :font))
  (:module rsl-games ("gaps-game" "spider-game"))
  (:module muffy-games ("baker's-dozen-game" "canfield-game" "klondike-game"))
  (:module muffy-games-2 ("calculation-game"))
  (:serial "card-definitions"
           "dw-patch"                    ; Adds dw:with-output-recording-disabled
           (:parallel "card-presentation-types" "card-places" "card-table" font)
           (:serial rsl-games muffy-games muffy-games-2)))
```

lisp-lore:examples:card-game:card-definitions.lisp

```
;;; -*- Mode: LISP; Syntax: Common-Lisp; Base: 10; Package: CARDS -*-
(defflavor card
        (rank
         suit
         glyph
         color
         (draw-alu)
         (overlay-alu)
         )
        ()
   (:initable-instance-variables rank suit)
   (:readable-instance-variables rank suit)
   (:required-init-keywords :rank :suit))

;;; Font size
(defparameter *card-width* 54)
(defparameter *card-height* 74)

(defparameter *suits* #(:spade :heart :diamond :club))

(defparameter *card-glyph-info*
              '((:Spade . #.(char-code #\A)) (:Heart . #.(char-code #\n))
                (:Diamond . #.(char-code #\N)) (:Club . #.(char-code #\a))))

(defparameter *empty-place-glyph-info*
              '((:Black #.(char-code #\0) :white) (:White #.(char-code #\1) :white)
                (:Joker #.(char-code #\2) :white)
                (:Card-Back #.(char-code #\3) :blue)
                (:fancy-card-back #.(char-code #\4) :blue)
                (:funny-joker #.(char-code #\?) :white)))

(defparameter *black-glyph* (cadr (assoc :black *empty-place-glyph-info*)))

(defparameter *rank-names*
              #("Ace" "Deuce" "Trey" "Four" "Five" "Six" "Seven"
                "Eight" "Nine" "Ten" "Jack" "Queen" "King"))

(defmethod (make-instance card) (&rest ignore)
  (assert (≤ 1 rank 13) ()
          "Your card must be an ace, a number between 2 and 10, or a jack, queen or king.")
  (assert (find suit *suits*) ()
          "Your suit must be Spade, Heart, Diamond or Club")
  (setf glyph (+ (cdr (assoc suit *card-glyph-info*)) rank -1))
  (setf color (case suit
                ((:spade :club) :black)
                ((:heart :diamond) :red))))

(defmethod (sys:print-self card) (stream ignore print-readably)
  (if print-readably
      (sys:printing-random-object (self stream) (princ self stream))
      (format stream "~A of ~:(~A~)s" (aref *rank-names* (1- rank)) suit)))
```

```
;;;
;;; Definitions for drawing (including color drawing) of cards.
;;;

;;; What kind of screen is it?  Possible values: nil, :dependent,
;;; :independent mean B&W, 8-bit color, 24-bit color, respectively.
(defun color-screen-map-mode (screen)
  (multiple-value-bind (nil nil mode) (send screen :color-map-description) mode))

;;; Cache for color ALU's -- only calculate new ones.
(defvar *color-alu-values* nil)

;;; Get color ALU for a given symbolic color.
(defun color-alu (screen alu color &optional (mask -1))
  (when color
    (unless *color-alu-values* (setup-colors screen))
    (or (getf *color-alu-values* color)
        (setf (getf *color-alu-values* color)
              (let ((alu (send screen :compute-color-alu alu color)))
                (send alu :set-plane-mask mask)
                alu)))))

;;; Add symbolic names for non-standard colors we will use.
(defun setup-colors (screen)
  (send screen :name-color 'dark-green 0 .4 .1))

;;; Return a color ALU or pattern for drawing backgrounds.  For
;;; example, the table background is either dark green or a gray
;;; pattern, depending on whether the screen is color or
;;; black-and-white.
(defun color-alu-or-pattern (stream alu color pattern &optional (mask -1))
  (declare (values alu pattern))
  (let* ((screen (send stream :screen))
         (mode (color-screen-map-mode screen)))
    (case mode
      ((nil) (values color:alu-x+y pattern))
      (:independent (values (color-alu screen alu color mask) t)))))

;;; Calculate the two alu's for cards.  The first one is either the
;;; :red or :black ALU.  The second one is the :yellow alu for
;;; picture cards only.
(defun-in-flavor (display-card-alus card) (stream)
  (let* ((screen (send stream :screen))
         (mode (color-screen-map-mode screen)))
    (case mode
      ((nil) (values color:alu-x+y nil))
      (:independent (values (color-alu screen color:alu-x color)
                            (color-alu screen color:alu-x*y
                                       (and (≤ 11 rank 13) :yellow)
                                       (lognot #o377))))
      (otherwise (error "We don't support ~A color screen maps yet." mode)))))

;;; Inside rectangle for the picture-card overlay
(defparameter *card-inside-left* 12)
(defparameter *card-inside-top* 5)
(defparameter *card-inside-right* 41)
(defparameter *card-inside-bottom* 69)

;;; For moving between the B&W and color screens, need to force
;;; recalculation of ALU's
(defmethod (clear-card-alu card) ()
  (setf draw-alu nil))
```

```
;;; Draw the card at the given spot.
(defmethod (display-card card) (stream x y)
  (unless draw-alu (multiple-value-setq (draw-alu overlay-alu) (display-card-alus stream)))
  (erase-card-spot stream x y)
  (graphics:draw-glyph glyph fonts:deck-of-cards x y :stream stream :alu draw-alu)
  (when overlay-alu
    (graphics:draw-rectangle (+ x *card-inside-left*) (+ y *card-inside-top*)
                             (+ x *card-inside-right*) (+ y *card-inside-bottom*)
                             :stream stream :alu overlay-alu)))

;;; Draw a background rectangle for a card.
(defun erase-card-spot (stream x y)
  (let* ((screen (send stream :screen))
         (mode (color-screen-map-mode screen)))
    (graphics:draw-glyph *black-glyph* fonts:deck-of-cards x y
                         :stream stream
                         :alu (ecase mode
                                ((nil) color:alu--x*y)
                                (:independent (color-alu screen color:alu-x :white))))))

;;; Draw special glyphs from font.  Used for card backs and jokers.
(defun display-empty-place (empty-appearance stream x y)
  (let* ((screen (send stream :screen))
         (mode (color-screen-map-mode screen))
         (invisible-p (eql empty-appearance :invisible))
         (glyph-data (unless invisible-p
                       (cdr (assoc empty-appearance *empty-place-glyph-info*))))
         (alu (if invisible-p color:alu-noop
                (ecase mode
                  ((nil) color:alu-x+y)
                  (:independent (color-alu screen color:alu-x (second glyph-data))))))
         (glyph (if invisible-p *black-glyph* (first glyph-data))))
    (unless invisible-p (erase-card-spot stream x y))
    (graphics:draw-glyph glyph fonts:deck-of-cards x y :stream stream :alu alu)))
```

```
;;;
;;; Card decks.  The array part is the deck.  next is for dealing.
;;;
(defstruct (card-deck
               (:type :named-array-leader)
               (:make-array (:length n-cards))
               (:copier nil)
               (:size-symbol card-deck-leader-length))
   n-cards
   (next 0))

;;; Make a card deck out of (* n-decks 52) cards.
(defun make-deck (&optional (n-decks 1))
   (let* ((n-cards (* n-decks 52))
          (deck (make-card-deck :n-cards n-cards)))
     (loop for deck-looper from 0 by 52 below n-cards
           do (loop for suit being the array-elements of *suits*
                    for suit-looper from 0 by 13
                    do (loop for number from 1 to 13
                             do (setf (aref deck (+ deck-looper suit-looper number -1))
                                      (make-instance 'card :suit suit :rank number)))))
     deck))

;;; Shuffle a deck.  The incoming deck is any kind of sequence.  The
;;; result is an array.  If you pass in an old deck, you can have
;;; cards shuffled into it.
;;; Note: the passed-in deck might actually be a list of cards: GAPS uses this.
(defun shuffle-deck (deck
                     &optional to-deck
                     &aux (n-cards (length deck)))
   (let ((n-cards (length deck)))
     (unless to-deck (setq to-deck (make-card-deck :n-cards n-cards)))
     (sys:with-stack-array (indices n-cards)
       (declare (sys:array-register indices))
       (loop for i from 0 below n-cards do (setf (aref indices i) i))
       (loop for i from 0 below (1- n-cards)
             as random = (random (- n-cards i)) do (rotatef (aref indices i)
                                                            (aref indices (+ random i)))))
     (loop for i from 0 below n-cards
           do (setf (aref to-deck i) (elt deck (aref indices i)))))))
   (setf (card-deck-next to-deck) 0)
   to-deck)

;;; Get next card out of the deck.
(defun deal-card (deck)
   (when (= (card-deck-next deck) (card-deck-n-cards deck))
     (error "Can't deal any more cards from this deck."))
   (prog1 (aref deck (card-deck-next deck))
          (incf (card-deck-next deck))))

;;; A macro to shuffle a sequence of cards into a temporary deck.
(defmacro with-cards-shuffled ((shuffled-cards cards) &body body)
   `(sys:with-stack-array (,shuffled-cards (length ,cards)
                                           :named-structure-symbol 'card-deck
                                           :leader-length card-deck-leader-length)
      (declare (sys:array-register ,shuffled-cards))
      (shuffle-deck ,cards ,shuffled-cards)
      ,@body))
```

lisp-lore:examples:card-game:card-presentation-types.lisp

```lisp
;;; -*- Mode: LISP; Syntax: Common-Lisp; Base: 10; Package: CARDS -*-

(defparameter *suit-translations* '(:spades :spade :hearts :heart
                                     :diamonds :diamond :clubs :club))

(defparameter *suits-presentation-type* '((member ,@*suit-translations*)))

(defparameter *rank-translations* '(:ace 1 :deuce 2 :two 2 :trey 3 :three 3 :four 4 :five 5
                                     :six 6 :seven 7 :eight 8 :nine 9 :ten 10
                                     :jack 11 :queen 12 :king 13))

(defparameter *ranks-presentation-type*
              '((member :ace 2 :deuce :two 3 :trey :three 4 :four
                        5 :five 6 :six 7 :seven 8 :eight 9 :nine
                        10 :ten :jack :queen :king)))

(define-presentation-type card (() &key deck)
  :no-deftype t
  :parser ((stream)
            (flet ((find-card (suit rank)
                     (let ((rank (or (getf *rank-translations* rank) rank))
                           (suit (or (getf *suit-translations* suit) suit)))
                       (if (not deck)
                           (make-instance 'card :rank rank :suit suit)
                           (or (find nil deck :test (lambda (ignore card)
                                                      (and (eql (card-suit card) suit)
                                                           (eql (card-rank card) rank))))
                               (cerror "Create ~:(~A ~A~) -- for debugging only."
                                       "Can't find ~:(~A ~A~) in deck ~S"
                                       suit rank)
                               (make-instance 'card :rank rank :suit suit)))))
                   (parse-card-description (card-description)
                     (cond ((= (length card-description) 3)
                            (destructuring-bind (rank of suit) card-description
                              (ignore of)
                              (values suit rank)))
                           ((member (first card-description) *suit-translations*)
                            (destructuring-bind (suit rank) card-description
                              (values suit rank)))
                           (t (destructuring-bind (rank suit) card-description
                                (values suit rank))))))
              (let ((card-description
                      (accept '((or ((sequence-enumerated
                                       ,*ranks-presentation-type*
                                       ((alist-member :alist (("of" . :of))))
                                       ,*suits-presentation-type*)
                                      :sequence-delimiter #\space :echo-space nil)
                                     ((sequence-enumerated
                                       ,*ranks-presentation-type*
                                       ,*suits-presentation-type*)
                                      :sequence-delimiter #\space :echo-space nil)
                                     ((sequence-enumerated
                                       ,*suits-presentation-type*
                                       ,*ranks-presentation-type*)
                                      :sequence-delimiter #\space :echo-space nil)))
                              :stream stream :prompt nil)))
                (multiple-value-bind (suit rank)
                    (parse-card-description card-description)
                  (find-card suit rank))))))
```

```
;;; Continuation of (define-presentation-type card (() &key deck)
    :describer ((stream &key plural-count)
                (cond ((null plural-count)
                       (princ "a card, rank of suit or suit rank" stream))
                      ((eql plural-count t)
                       (princ "cards in the form rank of suit or suit rank" stream))
                      ((numberp plural-count)
                       (format stream "~R cards in the form rank of suit or suit rank"
                               plural-count))
                      (t (format stream "~A cards in the form rank of suit or suit rank"
                                 plural-count)))))


;;; Presentation types for things you can only click on, not type:
;;; card-places and empty-places.

(define-presentation-type card-place ((&key empty-p))
     :no-deftype t
     :parser ((stream) (loop (dw:read-char-for-accept stream)))
     :typep ((place) (if empty-p (null (place-contents place)) t)))

(define-presentation-type empty-place (())
   :no-deftype t
   :parser ((stream) (loop (dw:read-char-for-accept stream))))
```

lisp-lore:examples:card-game:card-places.lisp

```
;;; -*- Syntax: Common-Lisp; Base: 10; Mode: Lisp; Package: CARDS -*-


;;; Error flavors
(defflavor place-error
        (place)
        (error)
  :initable-instance-variables)

(defflavor place-contents-may-not-be-removed
        ()
        (place-error))

(defmethod (dbg:report place-contents-may-not-be-removed) (stream)
  (format stream "The cards in ~S may not be removed. ~%They are:~{ ~A~^, ~}"
          place (place-contents place)))

(defflavor place-contents-may-not-be-stored
        (contents)
        (place-error)
  :initable-instance-variables)

(defmethod (dbg:report place-contents-may-not-be-stored) (stream)
  (format stream "The cards ~{ ~A~^, ~} may not be stored in ~S" contents place))


(compile-flavor-methods place-contents-may-not-be-removed place-contents-may-not-be-stored)


;;; Required in order to compile (setf (place-contents ...) ...) before
;;; any flavor methods have been defined yet.
(defgeneric (setf place-contents) (place new-contents))

;;; The raison d'etre for this stuff
(defun move-contents (from-place to-place)
  (unless (contents-may-be-removed from-place)
    (error 'place-contents-may-not-be-removed :place from-place))
  (let ((contents (place-contents from-place)))
    (unless (contents-may-be-stored to-place contents)
      (error 'place-contents-may-not-be-stored :place to-place :contents contents))
    (setf (place-contents to-place) contents
          (place-contents from-place) nil)))
```

```
;;; Like setf, only returns the first value instead of the last.
(defmacro setf1 (reference1 value1 &rest more-pairs)
  `(prog1 (setf ,reference1 ,value1)
          (setf ,@more-pairs)))

;;; There are superiors whose caches must also be invalidated.  This is for them.
(defgeneric invalidate-caches (place)
  (:method-combination :progn))

;;; Encaching mixins -- for runtime and consing efficiency
(defflavor contents-remembering-mixin
        (remembered-contents
         (remembered-contents-p))
        ()
  :abstract-flavor
  (:required-methods place-contents (setf place-contents)))

;;; Use setf1 to return value of whopper continuation (see definition above)
(defwhopper (place-contents contents-remembering-mixin) ()
  (if remembered-contents-p
      remembered-contents
      (setf1 remembered-contents (continue-whopper)
             remembered-contents-p t)))

(defwhopper ((setf place-contents) contents-remembering-mixin) (new-contents)
  ;; Don't encache new-contents.
  ;; Allow underlying flavor to be responsible for what the place contains.
  (setf remembered-contents-p nil)
  (continue-whopper new-contents))

(defmethod (invalidate-caches contents-remembering-mixin) ()
  (setf remembered-contents-p nil))


(defflavor rules-remembering-mixin
        (removal-rule-result
         (removal-rule-result-stored-p)
         store-rule-result
         (store-rule-result-stored-p))
        ()
  :abstract-flavor
  (:required-methods (setf place-contents) contents-may-be-removed contents-may-be-stored))

(defwhopper ((setf place-contents) rules-remembering-mixin) (new-value)
  (setf removal-rule-result-stored-p nil store-rule-result-stored-p nil)
  (continue-whopper new-value))

(defmethod (invalidate-caches rules-remembering-mixin) ()
  (setf removal-rule-result-stored-p nil store-rule-result-stored-p nil))

;;; Use setf1 to return value of whopper continuation (see definition above)
(defwhopper (contents-may-be-removed rules-remembering-mixin) ()
  (if removal-rule-result-stored-p
      removal-rule-result
      (setf1 removal-rule-result (continue-whopper)
             removal-rule-result-stored-p t)))

(defwhopper (contents-may-be-stored rules-remembering-mixin) (new-contents)
  (if (eql store-rule-result-stored-p new-contents)
      store-rule-result
      (setf1 store-rule-result (continue-whopper new-contents)
             store-rule-result-stored-p new-contents)))
```

```lisp
(defflavor self-presenting-mixin
        ((presentation)
         (empty-presentation)
         (window)
         (presented-at-x)
         (presented-at-y))
        ())

(defmethod (present-self self-presenting-mixin) (stream x y)
  (present-contents self stream x y)
  (present-empty-spaces self stream x y))

;;; Use setf1 to return value of continuation upward.
(defwhopper (present-contents self-presenting-mixin) (stream x y)
  (setf1 presentation
         (dw:with-output-as-presentation (:stream stream :object self
                                          :type 'card-place :single-box t)
           (continue-whopper stream x y))
         window stream presented-at-x x presented-at-y y))

(defwhopper (present-empty-spaces self-presenting-mixin) (stream x y)
  (setf1 empty-presentation
         (dw:with-output-as-presentation (:stream stream :object self :type 'empty-place
                                          :single-box t)
           (continue-whopper stream x y))))

(defmethod (erase-self self-presenting-mixin) (&optional recursion?)
  (macrolet ((erase-presentation (presentation)
               '(when ,presentation
                  (let ((boxes (dw::presentation-mouse-sensitive-boxes ,presentation window)))
                    (send window :delete-displayed-presentation ,presentation)
                    (loop for (left top right bottom) in boxes
                          do (draw-background-rectangle
                               window (- left 5) (- top 5) right bottom)))
                  (setf ,presentation nil))))
    (erase-presentation presentation)
    (erase-presentation empty-presentation)
    (unless recursion?
      (let ((superior (place-superior self)))
        (when superior (erase-self superior t))))))

(defmethod (redisplay-self self-presenting-mixin) ()
  (when presentation
    (erase-self self)
    (present-self self window presented-at-x presented-at-y)))

(defmethod (highlight-self self-presenting-mixin) ()
  (when presentation
    (loop for (left top right bottom)
          in (dw::presentation-mouse-sensitive-boxes presentation window)
          do (draw-highlighting-rectangle window (- left 5) (- top 5) right bottom))))

(defmethod (unhighlight-self self-presenting-mixin) ()
  (redisplay-self self))

(defmethod ((setf place-contents) self-presenting-mixin :after) (ignore)
  (loop for place = self then (place-superior place)
        while (place-superior place)
        finally (redisplay place)))
```

```
;;; A place to put cards.  Implements the rules supplied by a game.
(defflavor basic-card-place
        ((name (gensym))
         superior
         (game (card-table-game dw:*program*))
         (addition-rules)
         (removal-rules)
         (addition-side-effects)
         (removal-side-effects))
        (contents-remembering-mixin rules-remembering-mixin self-presenting-mixin)
  :abstract-flavor
  (:conc-name place-)
  (:init-keywords :contents)
  (:readable-instance-variables name superior)
  :initable-instance-variables
  (:required-init-keywords :superior :game)
  (:required-methods present-self place-contents (setf place-contents)))


(defvar *no-side-effects* nil)

;;; Make sure the contents of a place get stored in it.
(defmethod (make-instance basic-card-place) (&key contents &allow-other-keys)
  (let ((*no-side-effects* t))
    (setf (place-contents self) contents)))

;;; Implement the removal rule.
(defmethod (contents-may-be-removed basic-card-place) ()
  (and (place-contents self)
       (catch 'forbidden
         (loop for rule in removal-rules
               thereis (funcall rule game self (place-contents self)))))))

;;; Implement the store rule.
(defmethod (contents-may-be-stored basic-card-place) (new-contents)
  (when (not (place-contents self))
    (catch 'forbidden
      (loop for rule in addition-rules
            thereis (funcall rule game self new-contents)))))

(defmethod ((setf place-contents) basic-card-place :after) (new-contents)
  (unless *no-side-effects*
    (loop for side-effect in (if new-contents addition-side-effects removal-side-effects)
          do (funcall side-effect game self))))
```

```
;;; A place that holds exactly one card.
(defflavor card-place
        ((card)
         empty-appearance)
        (basic-card-place)
  (:initable-instance-variables empty-appearance)
  (:required-init-keywords :empty-appearance))

;;; Get back the cards in the place (all one of them).
(defmethod (place-contents card-place) ()
  (and card (list card)))

;;; Put a card into its place.
(defmethod ((setf place-contents) card-place) (new-contents)
  (setf card (first new-contents)))

;;; A card can only be stored into an empty card-place.  Only one card to a place.
(defwhopper (contents-may-be-stored card-place) (new-contents)
  (and (null card)
       (null (cdr new-contents))
       (continue-whopper new-contents)))


;;; A card place displays itself
(defmethod (present-contents card-place) (stream x y)
  (when card
       (display-card card stream x y)))

(defmethod (present-empty-spaces card-place) (stream x y)
  (unless card
     (display-empty-place empty-appearance stream x y)))
```

```lisp
;;; A stack of cards.  Holds zero cards, one card, or a card plus a stack of them.
(defflavor stack-place
           ((card)
            (rest-of-stack)
            delta-x
            delta-y
            empty-appearance)
           (basic-card-place)
  (:readable-instance-variables card rest-of-stack)
  (:initable-instance-variables delta-x delta-y empty-appearance)
  (:required-init-keywords :delta-x :delta-y :empty-appearance))


;;; Get the contents of a stack place.
(defmethod (place-contents stack-place) ()
  (and card
       (list* card (if rest-of-stack (place-contents rest-of-stack)
                       (error "No rest of stack behind card??")))))

;;; Store new contents into a stack.
(defmethod ((setf place-contents) stack-place) (new-contents)
  (setf card (first new-contents)
        rest-of-stack
        (and card
             (make-instance 'stack-place :contents (rest new-contents)
                                         :superior self
                                         :game game
                                         :addition-rules addition-rules
                                         :removal-rules removal-rules
                                         :addition-side-effects addition-side-effects
                                         :removal-side-effects removal-side-effects
                                         :delta-x delta-x :delta-y delta-y
                                         :empty-appearance empty-appearance)))
  (loop for superior = self then (place-superior superior)
        while superior
        do (invalidate-caches superior))
  (place-contents self))

(defmethod (present-contents stack-place) (stream x y)
  (when card
    (display-card card stream x y)
    (when rest-of-stack
      (present-contents rest-of-stack stream (+ x delta-x) (+ y delta-y)))))

(defmethod (present-empty-spaces stack-place) (stream x y)
  (if card
      (when rest-of-stack
        (present-empty-spaces rest-of-stack stream (+ x delta-x) (+ y delta-y)))
      (display-empty-place empty-appearance stream x y)))

(defwhopper (erase-self stack-place) (&optional recursion?)
  (when rest-of-stack (erase-self rest-of-stack t))
  (continue-whopper recursion?))
```

```
(defflavor stack-with-face-down-cards-place
        ((face-down-cards))
        (stack-place)
  :initable-instance-variables)

(defwhopper ((setf place-contents) stack-with-face-down-cards-place) (new-contents)
  (if new-contents
      (continue-whopper new-contents)
      (redisplay self)
      (continue-whopper (and face-down-cards (list (pop face-down-cards)))))))

(defwhopper (present-self stack-with-face-down-cards-place) (stream x y)
  (if face-down-cards
    (display-empty-place :fancy-card-back stream (- x delta-x) (- y delta-y))
    (let ((x (- x delta-x)) (y (- y delta-y)))
      (display-empty-place :black stream x y)))
  (sys:letf-if (not card) ((empty-appearance :white))
      (continue-whopper stream x y)))


(compile-flavor-methods card-place stack-place stack-with-face-down-cards-place)
```

lisp-lore:examples:card-game:card-table.lisp

```
;;; -*- Mode: LISP; Syntax: Common-Lisp; Base: 10; Package: CARDS -*-

(dw:define-program-framework card-table
  :select-key #\Square
  :command-definer t
  :command-table (:kbd-accelerator-p t
                  :inherit-from '())
  :top-level (dw:default-command-top-level :echo-stream ignore)
  :state-variables ((game))
  :panes
  ((board :display :typeout-window t
          :redisplay-after-commands t
          :incremental-redisplay :own-redisplayer
          :redisplay-function 'display-game-board)
   (title :title :redisplay-string "Card Games" :height-in-lines 1
    :redisplay-after-commands nil)
   (input :interactor :height-in-lines 8)
   (games :command-menu :center-p t :columns 1 :menu-level :games)
   (commands :command-menu :center-p t :columns 1 :menu-level :top-level))
  :configurations
  '((main
     (:layout (main :column board title row-1)
      (row-1 :row input games commands))
     (:sizes (main (title 1 :lines) :then (row-1 8 :lines input) :then (board :even))
      (row-1
       (games :ask-window self :size-for-pane games)
       (commands :ask-window self :size-for-pane commands) :then
       (input :even))))))

(defmacro current-game ()
  '(card-table-game dw:*program*))

(defmacro-in-flavor (current-game card-table) ()
  'game)

(defmacro place-to-move-valid ()
  '(and (boundp 'dw:*program*) (typep dw:*program* 'card-table) (current-game)))

(defmacro place-to-move ()
  '(game-place-to-move (current-game)))

(defmacro card-table () '(dw:get-program-pane 'board))

(defun complete-redisplay () (redisplay :complete-redisplay))

(defvar *redisplay-list* nil)

(defun redisplay (place) (pushnew place *redisplay-list*))

(defmethod (display-game-board card-table) (stream)
  (if (or (null *redisplay-list*)                    ; ?? Unsure this is the right thing.
          (member :complete-redisplay *redisplay-list*))
      (display-whole-game-board self stream)
      (loop for place in (nreverse (prog1 *redisplay-list* (setf *redisplay-list* nil)))
            do (redisplay-self place))))

(defmethod (display-whole-game-board card-table) (stream)
  (setq *redisplay-list* nil)
  (send stream :clear-history nil t)
  (multiple-value-bind (left top right bottom) (send stream :inside-edges)
    (draw-background-rectangle stream left top right bottom))
  (when game
    (display-game game stream)))
```

```
(defun draw-background-rectangle (stream left top right bottom)
  (dw:with-output-recording-disabled (stream)
    (graphics:draw-rectangle left top right bottom :stream stream :filled t :alu tv:alu-setz)
    (multiple-value-bind (alu pattern)
        (color-alu-or-pattern stream color:alu-x 'dark-green tv:10%-gray)
      (graphics:draw-rectangle left top right bottom :stream stream
                                                      :filled t :alu alu :pattern pattern))))

(defun draw-highlighting-rectangle (stream left top right bottom)
  (dw:with-output-recording-disabled (stream)
    (multiple-value-bind (alu pattern)
        (color-alu-or-pattern stream color:alu-x*y :cyan tv:25%-gray)
      (graphics:draw-rectangle left top right bottom :stream stream
                                                     :filled t :alu alu :pattern pattern))))

(defflavor basic-game
        ((name)
         (card-places)
         (card-place-descriptions)
         (raw-deck)
         (shuffled-deck)
         (place-to-move))
        ()
  (:initable-instance-variables card-place-descriptions name)
  (:writable-instance-variables place-to-move)
  (:conc-name game-)
  (:method-combination start-new-game (:progn :most-specific-last)
                       make-card-place-descriptions (:append :most-specific-first)))

(defmacro-in-flavor (current-game basic-game) ()
  'self)

(defmethod (make-instance basic-game) (&rest ignore)
  (unless  name (setf name (string-capitalize-words (type-of self))))
  (unless card-place-descriptions
    (setf card-place-descriptions
          (make-card-place-descriptions self (dw:get-program-pane 'board)))))

;;; Use #. to calculate constant at compile time.
(defun-in-flavor (divide-up-space basic-game) (stream n-columns &optional n-rows)
  (incf n-columns) (when n-rows (incf n-rows))
  (multiple-value-bind (width height) (send stream :inside-size)
    (let* ((column-width (round width (1+ n-columns)))
           (column-extra column-width)
           (row-height (and n-rows (round height (1+ n-rows))))
           (row-extra row-height))
      (when (< column-width #.(round (* *card-width* 6/5)))
        (setf column-width (round width n-columns) column-extra (round column-width 2)))
      (when (and n-rows (< row-height #.(round (* *card-height* 6/5))))
        (setf row-height (round height n-rows) row-extra (round row-height 2)))
      (values column-extra column-width row-extra row-height))))
```

```
(defmethod (start-new-game basic-game) ()
  (unless raw-deck (setf raw-deck (make-deck (game-n-decks self))))
  (setf shuffled-deck (setf shuffled-deck (shuffle-deck raw-deck shuffled-deck)))
  (setf card-places (make-card-places self)))

(defmethod (game-n-decks basic-game :default) () 1)        ; By default, all games use one deck.

(defmethod (display-game basic-game) (stream)
  (loop for (nil (x y)) in card-place-descriptions
        and place in card-places
        do (present-self place stream x y))
  (when (place-to-move)
    (highlight-self (place-to-move))))

(defmethod (make-card-places basic-game) ()
  (loop with *no-side-effects* = t
        for (place-type () . keywords) in card-place-descriptions
        collect (apply #'make-card-place self place-type keywords)
        finally (complete-redisplay)))

(defgeneric make-card-place (game place-type &rest place-keywords)
  (declare (values card-place))
  (:method-combination :case))

(defmethod (make-card-place basic-game :card)
           (&rest options &key n-cards &allow-other-keys)
  (si:with-rem-keywords (make-instance-options options '(:n-cards))
    (apply #'make-instance 'card-place
           :superior nil
           :game (current-game)
           :contents (when (and n-cards (plusp n-cards))
                       (if (= n-cards 1)
                           (list (deal-card shuffled-deck))
                           (error "n-cards ≠ 1?")))
           make-instance-options)))

(defmethod (make-card-place basic-game :stack)
           (&rest options &key n-cards &allow-other-keys)
  (si:with-rem-keywords (make-instance-options options '(:n-cards))
    (apply #'make-instance 'stack-place
           :superior nil
           :game (current-game)
           :contents (when n-cards
                       (loop repeat n-cards
                             collect (deal-card shuffled-deck)))
           make-instance-options)))

(defmethod (make-card-place basic-game :stack-with-face-down-cards)
           (&rest options &key n-cards n-face-down &allow-other-keys)
  (si:with-rem-keywords (make-instance-options options '(:n-cards :n-face-down))
    (apply #'make-instance 'stack-with-face-down-cards-place
           :superior nil
           :game (current-game)
           :contents (when n-cards
                       (loop repeat n-cards
                             collect (deal-card shuffled-deck)))
           :face-down-cards (when n-face-down
                              (loop repeat n-face-down
                                    collect (deal-card shuffled-deck)))
           make-instance-options)))
```

```
(dw:define-presentation-action move-card-supply-from
   (card-place cp:command
    :tester ((place) (and (place-to-move-valid) (not (place-to-move))
                          (contents-may-be-removed place)))
     :documentation ((place) (format nil "Move~{ ~A~^,~}" (place-contents place))))
   (from-place)
   (setf (place-to-move) from-place)
   (highlight-self from-place)
   )

(define-presentation-to-command-translator move-card-supply-to
   (empty-place
    :tester ((place) (and (place-to-move-valid) (place-to-move)
                          (contents-may-be-stored place (place-contents (place-to-move)))))
     :documentation ((ignore) (format nil "Move~{ ~A~^,~}" (place-contents (place-to-move)))))
   (to-place)
   `(com-move-contents-end ,to-place))

(define-card-table-command (com-move-contents-end :name nil) ((to 'card-place))
   (move-contents (place-to-move) to)
   (setf (place-to-move) nil))

(define-card-table-command (com-oops) ()
   (let ((place (place-to-move)))
     (when place
       (loop for place = place then (place-superior place)
             while (place-superior place)
             finally (redisplay place)))
     (setf (place-to-move) nil)))

(define-presentation-to-command-translator oops-my-dear
   (t
    :gesture :middle
    :tester ((ignore) (and (place-to-move-valid) (place-to-move)))
    :documentation ((ignore) (format nil "Abort current attempted move of~{ ~A~^,~}"
                               (place-contents (place-to-move))))
    :blank-area t :suppress-highlighting t)
   (ignore)
   `(com-oops))

(define-card-table-command (com-start-new-game :menu-accelerator "New Game") ()
   (when game
     (start-new-game game)
     (cp::command-table-update-options (cp:find-command-table "Card-Table")
                                       :inherit-from (list (command-table-name game)))
     (complete-redisplay))
   (setf (place-to-move) nil))

(define-card-table-command (com-show-card-places :keyboard-accelerator #\?) ()
   (when game (show-card-places game)))

(define-card-table-command (com-next-round :keyboard-accelerator #\complete
                                           :menu-accelerator "Next Round") ()
   (when game (start-next-round game)))

(defmethod (start-next-round basic-game :default) ()
   (format t "~2%I don't know how to start a new round for a ~A.~2%"
           (string-capitalize-words (type-of self))))
```

```
(define-card-table-command (com-color-screen :menu-accelerator "Color") ()
   (locally
       (declare (special color:color-screen))        ; Get rid of warning.
       (if (send (send dw:*program-frame* :screen) :color-map-description)
           (format t "~2%Already on the color screen.~2%")
           (multiple-value-bind (type memory) (color:color-system-description)
               (cond ((not type)
                         (with-character-style ('(nil :bold :very-large))
                             (format t "~2%Sorry, no color system on this machine.~2%")))
                     ((< memory 3)
                         (with-character-style ('(nil :bold :very-large))
                             (format t "~2%Sorry, this machine does not have 24-bit color.")))
                     (t (when (or (not (variable-boundp color:color-screen))
                                  (null color:color-screen))
                             (setf color:color-screen
                                   (funcall 'color:make-color-screen    ; Get rid of compiler warning
                                            :setup :standard))
                             (send color:color-screen :expose))
                        (send dw:*program-frame* :set-save-bits nil)
                        (send dw:*program-frame* :set-superior color:color-screen)
                        (when game (clear-card-alus game))
                        (tv:mouse-set-sheet color:color-screen)))))))

(define-card-table-command (com-b&w-screen :menu-accelerator "B&W") ()
   (if (not (send (send dw:*program-frame* :screen) :color-map-description))
       (format t "~2%Already on the black-and-white screen.~2%")
       (send dw:*program-frame* :set-superior tv:main-screen)
       (tv:mouse-set-sheet tv:main-screen)
       (when game (clear-card-alus game))
       (send dw:*program-frame* :set-save-bits t)))

(defmethod (clear-card-alus basic-game) ()
   (loop for index below (card-deck-n-cards shuffled-deck)
         do (clear-card-alu (aref shuffled-deck index))))

(defmethod (show-card-places basic-game) ()
   (loop for card-place in card-places
         do (format t "~%~S contains: ~{~A~^, ~}" card-place (place-contents card-place))))
```

```lisp
(defprop define-game "Game" si:definition-type-name)

(defmacro define-game (name instance-variables &optional flavors-built-on &rest options)
  (when (record-source-file-name name 'define-game)
    (let* ((game-name (string-capitalize-words name))
           (comtab-name (string-append game-name " Command Table"))
           (command-definer-name (intern (string-append "DEFINE-" name "-COMMAND")
                                          (symbol-package name)))
           (command-name (intern (string-append "COM-" name) (symbol-package name))))
      `(progn
         (add-initialization ,(format nil "Create ~A" comtab-name)
                             '(cp:make-command-table
                                ',comtab-name
                                :inherit-from '("Colon Full Command"
                                                "Standard Arguments" "Standard Scrolling"))
                             '(:once))

         (si:defflavor-with-parent (,name define-game) ,name
                                   ,instance-variables
                                   (,@flavors-built-on basic-game)
           :writable-instance-variables
           ,@options
           (:default-init-plist :name ,game-name))

         (defmacro ,command-definer-name (command-name-and-options args &body body)
           (let* ((command-name-with-options (if (listp command-name-and-options)
                                                 command-name-and-options
                                                 (list command-name-and-options)))
                  (command-name (first command-name-with-options))
                  (real-command-name (intern (string-append ',name "-" command-name)
                                             (symbol-package command-name)))
                  (command-options (rest command-name-with-options))
                  (command-name-string
                    (let ((command-name-string-temp (string-capitalize-words command-name)))
                      (if (string-equal command-name-string-temp "Com "
                                        :end1 4 :end2 4)
                          (substring command-name-string-temp 4)
                          command-name-string-temp)))
                  (arguments (loop for (name) in args collect name))
                  (internal-name (intern (string-append command-name "-COMMAND-BODY")
                                         (symbol-package command-name))))
             `(progn (define-card-table-command
                       (,real-command-name :command-table ,,comtab-name
                        :menu-accelerator ,command-name-string
                        :name ,command-name-string
                        ,@command-options)
                       ,args
                       (declare (sys:function-parent ,',name define-game))
                       (,internal-name game ,@arguments))
                     (defmethod (,internal-name ,',name) (,@arguments)
                       ,@body))))

         (defmethod (command-table-name ,name) () ',comtab-name)

         (define-card-table-command (,command-name :menu-accelerator ,game-name
                                     :menu-level :games)
           ()
           (declare (sys:function-parent ,name define-game))
           (setf game (make-instance ',name))
           (com-start-new-game))))))

(compile-flavor-methods card-table)
```

lisp-lore:examples:card-game:gaps-game.lisp

```
;;; -*- Mode: LISP; Syntax: Common-Lisp; Base: 10; Package: CARDS -*-

(define-game gaps-game (aces (game-in-progress)))

(defmethod (make-card-place-descriptions gaps-game :append) (stream)
  (multiple-value-bind (start-left column-width start-top row-height)
      (divide-up-space stream 13 4)
    (loop for y from start-top by row-height
          for row from 1 to 4
          nconc (loop for x from start-left by column-width
                      for column from 1 to 13
                      collect `(:card (,x ,y)
                                :n-cards 1
                                :addition-rules (,#'valid-gaps-move)
                                :removal-rules (,#'true)
                                :empty-appearance :invisible
                                :addition-side-effects (,#'check-for-win)
                                :name ,(+ (* row 13) column -14))))))

(defun-in-flavor (discard-aces gaps-game) ()
  (setf aces (loop with *no-side-effects* = t
                   for card-place in card-places repeat 52
                   as card = (first (place-contents card-place))
                   when (eql (card-rank card) 1)
                     collect card
                     and do (setf (place-contents card-place) nil))))

(defmethod (start-new-game gaps-game) ()
  (setf game-in-progress nil)
  (unless (= (card-deck-next shuffled-deck) 0)
    (setf shuffled-deck (shuffle-deck raw-deck shuffled-deck)))
  (loop with *no-side-effects* = t
        for card-place in card-places repeat 52
        as card = (deal-card shuffled-deck)
        do (setf (place-contents card-place) (list card)))
  (discard-aces)
  (complete-redisplay)
  (setf game-in-progress t))

(defmethod (valid-gaps-move gaps-game) (place new-contents)
  (let* ((new-card (first new-contents))
         (place-name (place-name place))
         (prev-place (and (plusp (mod place-name 13)) (elt card-places (1- place-name)))))
    ;; A move is valid iff:
    ;; 1. The card to its left is one less than the current one in the current suit, or
    ;; 2. The card is a deuce, and is being played in the leftmost column.
    (if prev-place
        (let ((prev-card (first (place-contents prev-place))))
          (and prev-card (eql (card-suit prev-card) (card-suit new-card))
               (= (card-rank new-card) (1+ (card-rank prev-card)))))
        (= (card-rank new-card) 2))))
```

```
(defmethod (winning-counts gaps-game) ()
  (loop repeat 4
        with all-places = card-places
        as row-card = (first (place-contents (pop all-places)))
        as row-suit = (and row-card (card-suit row-card))
        and row-rank = (and row-card (card-rank row-card))
        when (eql row-rank 2)
          collect (1+
                    (loop repeat 11
                          as card = (first (place-contents (pop all-places)))
                          as suit-to-check = (and card (card-suit card))
                          and rank-to-check = (and card (card-rank card))
                          when (and (eql suit-to-check row-suit)
                                    (eql rank-to-check (incf row-rank)))
                            count t
                          else do (setq row-suit nil)
                          finally (pop all-places)))
              else collect 0
                and do (loop repeat 12 do (pop all-places)))))

(defmethod (check-for-win gaps-game) (place)
  (ignore place)
  (when game-in-progress
    (let ((winning-ways (winning-counts self)))
      . (when (equal winning-ways '(12 12 12 12))
        (setf game-in-progress nil)
        (with-character-style ('(nil :bold :very-large) t :bind-line-height t)
          (format t "~3%You Win!  Click on \"New Game\" for another game.~2%"))
        (complete-redisplay)))))

(defmethod (start-next-round gaps-game) ()
  (let* ((winning-counts (winning-counts self))
         (losing-cards (loop for win-count in winning-counts
                             with all-places = card-places
                             do (loop repeat win-count do (pop all-places))
                                (append (loop repeat (- 13 win-count)
                                              collect
                                                (or (first (place-contents (pop all-places)))
                                                    (pop aces)))))))
    (with-cards-shuffled (new-shuffle losing-cards)
      (loop for win-count in winning-counts
            with all-places = card-places
            do (loop repeat win-count do (pop all-places))
               (loop repeat (- 13 win-count)
                     do (setf (place-contents (pop all-places))
                              (list (deal-card new-shuffle)))))))
  (discard-aces))
```

11.6 Problem Set

1. Implement a new game.

2. Simplify application of the rules by making them be methods on games instead of lists of functions to be called.

11.7 Fun and Games

CONS *(kahnz) verb.* To add a new element to a list, usually to the top rather than at the bottom.

CONS UP *verb.* To synthesize from smaller pieces; more generally, to create or intent. Examples: "I'm trying to cons up a list of volleyball players." "Let's cons up an example."
This term comes from the LISP programming language, which has a function called CONS that adds a data item to the front of a list.

12. More Advanced Use of the Editor

The standard Zwei and Zmacs[1] commands are generally quite well documented by the on-line help facilities, both within the Document Examiner and within Zmacs itself. Thus, there should be no difficulty in becoming fluent in the use of the built-in commands simply by consulting the automatic documentation. Or, if you prefer, many of the more common built-in commands are described on paper: See the section "Writing and Editing Code" in *Program Development Utilities*.

The methods for adding new commands, on the other hand, are not documented so completely. It is upon that topic that this chapter will concentrate.

---

[1]As I understand it, *eine* and *zwei*, apart from being "one" and "two" in German, were the names of the first two text editors written for MIT Lisp Machines. They are recursive acronyms, respectively, for Eine Is Not Emacs, and Zwei Was Eine Initially. Zwei is actually an editor substrate, used for implementing Zmail as well as Zmacs, which is the text editor proper.

12.1 Keyboard Macros

Keyboard macros allow you to bundle up any number of keystrokes and execute them all with one keystroke. (These actually are documented, but since they fit in with the rest of this chapter, I thought we should look at them as well.) The Zmacs command "c-X (" starts a keyboard macro. Whatever keys you press from then up until you type a "c-X)" are remembered while they are executed. When you type the c-X) the macro will be defined. It can be re-executed by typing c-X E. The effect will be as though you had typed all the keystrokes in the macro definition (but faster). Giving a numeric argument to c-X E will cause the macro to be repeated that many times.[2]

c-X E always executes the most recently defined macro, so if you define another macro with c-X (, the definition of the first one will be lost, unless you have previously saved it somehow. You can install it with m-X Install Macro, which will put it on a keystroke.[3] From then on (until you deinstall the macro, or install some other command on the same key), typing that keystroke will execute the macro.

Here's a simple example, something which I often did while working on the examples in this book. I'll define a keyboard macro for changing the lisp expression I just typed into a bold character style:

---

[2]You can also give the numeric argument to c-X), in which case it will be done one fewer than that number of times (you already did it once as you typed it in!)

[3]You can also name it, with m-X Name Last Kbd Macro, after which you can later install it with m-X Install Macro.

c-X (	start keyboard macro definition
c-Space	set the mark
c-m-B	go back one form
c-X c-X	swap point and mark[4]
c-X c-J	change style in region
B Return	to boldface
c-X)	finish macro definition

m-X Install Macro	
	prompted with "name of macro to install"
Return	Choose default, last one defined.
c-m-sh-J	"Key to get it"
Zmacs	"Install in what comtab:"[5]

The next step is to put something into my init file which would automatically define this keyboard macro every time I login. Here is one way to do it:

```
(zwei:define-keyboard-macro change-form-style-to-bold (nil)
  #\c-Space #\c-m-B #\c-x #\c-X #\c-X #\c-J #\B #\Return)

(zwei:command-store (zwei:make-macro-command
                        :change-form-style-to-bold)
                    #\c-m-sh-J
                    zwei:*zmacs-comtab*)
```

If I were adding several macros at one time, and to the same command table, I would use **zwei:set-comtab** rather than **zwei:command-store**. Command tables are discussed in the next section. See the section "Command Tables and Command Definition," page 286.

---

[4]Not required, but it puts the "point" back where it was before the command.

[5]I could also have clicked on "Zmacs."

12.2 Writing New Commands

Many extensions to the editor are not expressible as a sequence of keystrokes. For these, you need to write a function, with **zwei:defcom**, and then add it to the command table of your choice with **zwei:command-store** or **zwei:set-comtab**. Among the things you may want to do from your function are: insert text into a buffer, read text out of a buffer, get user input from the mini-buffer, and send text to the typein window. All of these are reasonably straightforward, once you know about a few key variables and functions.

12.2.1 Zwei Data Structure

The Zwei data structure consists primarily of four parts, of which I will describe three. The fourth, called a **window**, is used mostly by the redisplay.

12.2.1.1 Lines

A *line* is the basic Zwei unit of text. It is a string array with its leader used as a structure. The array itself holds the text of the line. The structured part of a line contains at least the following (accessors are listed in parentheses):

- The length of the line (**fill-pointer**)

- The lines immediately before and after this one (**zwei:line-previous** and **zwei:line-next**)

- The "time" at which this line was last updated (called a "tick") – used by the redisplay (**zwei:line-tick**)

12.2.1.2 BP's

A *BP* ("buffer pointer") is a pointer to a specific character in a line. BP's are lists, but their elements should be obtained with the following accessors:

zwei:bp-line   the line into which this BP points.

zwei:bp-index   the character offset into the line.

zwei:bp-status   what kind of BP this is: **:normal** means that it is an absolute location in the line. **:moves** means that if you insert or delete text in front of the BP, it will move accordingly.

zwei:bp-buffer the interval in which you can find this line.

Many operations are defined to take BP's as arguments, often returning other BP's as their results. For example, **zwei:forward-char** takes a BP and a numeric argument, and returns the BP which points to the character that many characters forward from the given one (negative means backwards). Other similar functions are **zwei:forward-line, zwei:forward-sentence, zwei:forward-paragraph** and **zwei:forward-sexp**.

Other functions you are probably going to be interested in include **zwei:move-bp** and **zwei:bp-<**. The first takes either a second BP or a line and index, and change the given BP to point to that location. The second takes two BP's and says whether the first occurs earlier in its buffer than the second.

12.2.1.3 Intervals

An *interval* is a pair of BP's which delimit the beginning and end of the interval. Depending on what the interval is used for, it might have other properties. For example, a file buffer is an interval which remembers its pathname and whatever is defined in the file; if it's a file which contains lisp code, the functions and variables declared in the file are remembered in sub-intervals called sections.[6]

One thing to notice is that an interval doesn't remember all the lines in it, just the first and last. To find out whether one line is before or after another is a potentially very slow process; **zwei:bp-<** is called as infrequently as possible. In general, functions which take two BP's as arguments take a third argument called **in-order-p**, which is true if the first one appears earlier than the second one in the interval. The macro **zwei:get-interval** takes these three variables and puts the two BP's in order. You can also pass an interval as the first BP, and **nil** as the second one, in which case you will get the BP's which refer to the beginning and end of the region.

12.2.1.4 Interval Streams

An easy way to read data from an interval, or add new data to an interval, is with a flavor of stream called an *interval stream*. The standard way to make an interval stream is with **zwei:open-editor-stream**, and its sidekick **zwei:with-editor-stream**. These open a bidirectional stream to an editor buffer. They are analogous to **open** and **with-open-file** in that **open-editor-stream** simply creates the stream and returns it, while **with-editor-stream** puts a call to **open-editor-stream** inside a useful wrapper, and so is preferable if your control structure allows it. (The wrapper in this case guarantees not a **zl:close**, which isn't meaningful for editor streams, but a **:force-redisplay**,) so any changes to the buffer will be apparent.)

There is some documentation on these two functions in chapter 45 of volume 7B, mainly on the various options for specifying which buffer the stream should point to, and where in the buff-

---

[6]Which is how m-. manages to find most definitions accurately.

er it should initially point. You must specify at least one of the following options: :interval, :buffer-name, :pathname, :window or :start. :buffer-name and :pathname are easy enough. If a buffer exists which matches the given information, it is used; if not, one is created (unless the :create-p option has been used to specify otherwise).

Any interval you might have your hands on is suitable for :interval. A convenient one is often the value of zwei:*interval*, which is valid in the editor process. You can also create an interval with zwei:make-interval, which takes a type (nil or a flavor built on zwei:interval) and the optional keyword :initial-line, which can be any string (including one with more than one line in it.)

:start can be a BP. It can also be :beginning, :end, :point, :mark, or :region. The last three are only valid if you pass in a :window argument.

:window is a zwei:window structure, not to be confused with objects of flavor tv:window. A Zwei window contains information used by the redisplay, about the portion of the buffer currently visible. Among its slots are a pointer to the interval that window is displaying part of, a BP for the position of *point* (the cursor), a bp for the first character in the line currently displayed at the top of the screen, and a user-movable BP which marks the "region" (called, unsurprisingly, the *"mark."*). The window structure you are most likely to be interested in is in the variable zwei:*window*, which is valid in the editor process.

The macro zwei:point, called with no arguments, returns the BP for the current point. As you might expect, it expands into (window-point *window*), which means it is only valid inside the editor. Similarly, there is a macro named zwei:mark, which returns a BP for the most recently dropped mark.

Here are a few trivial examples, to illustrate the basic concepts (all assume the current package is **zwei**).[7]

```
(with-editor-stream (stream :interval *interval*)
  (send stream :string-out
        "surprise text insterted at end of current buffer"))

(with-editor-stream (stream :interval *interval*
                               :start :beginning)
  (send stream :string-out
        "surprise inserted at beginning of current buffer"))

(with-editor-stream (stream :start (point))
  (send stream :string-out
        "surprise text inserted at point"))

(with-editor-stream (stream :start :region)
  (read stream))              ; returns form in region.

(with-editor-stream
  (stream :pathname "cd:>rsl>lispm-init.lisp"
          :load-p t
          :start :beginning)
  (send stream :line-in))    ; reads first line of indicated file
```

12.2.2 Command Tables and Command Definition

Writing new commands becomes a matter of figuring out what you want to do and expressing it in a function. Then, you need to hook the function up to the Zwei command table mechanism.

The macro for defining new commands is **zwei:defcom**. Unfor-

---

[7]Warning! **zwei** is a Zetalisp package, not common lisp. Many pitfalls await the unwary.

tunately, it has a slightly confusing syntax, but is otherwise not difficult. Its syntax is as follows:

```
(defcom com-foo documentation options-list &rest body)
```

This defines a function named **com-foo**[8]. The documentation is any string you want to print in response to Help requests, or the name of a function to print that help. The list of options can be selected from:

- **zwei:km** – This command preserves MARK if it is set (the default is to remove it).
- **zwei:sm** – This command sets MARK.
- **zwei:nm** – This command removes MARK.
- **zwei:r** – Recenter screen like c-N if moved off (positive arg means moving down)
- **zwei:-r** – Recenter screen like c-P if moved off (positive arg means moving up)
- **zwei:push** – Point is pushed on the point-pdl before executing

The body is the function which actually implements the command. Note that there is no lambda-list: the command is called with no arguments. Rather, arguments are passed in using special variables. **zwei:*numeric-arg*** is the prefix numeric argument typed by the user. **zwei:*numeric-arg-p*** tells whether a numeric argument was typed. **zwei:*last-command-char*** is the keystroke that was actually typed.

One last thing: your command body *must* return the value of one of the following variables to tell the redisplay what got changed:

---

[8]Zwei commands, like CP commands, are traditionally named **com-foo**, for some value of "foo." See hacker's definition at the end of the chapter.

- **zwei:dis-none** – No redisplay needed.
- **zwei:dis-mark-goes** – No redisplay needed, except maybe removing region underlining.
- **zwei:dis-bps** – Point and mark may have moved, but text is unchanged.
- **zwei:dis-line** – Text in one line may have changed.
- **zwei:dis-text** – Any text might have changed.
- **zwei:dis-all** – Global parameters of the window have changed. Clean the window and redisplay all lines from scratch.

To make a new command usable, you must put it into a command table. Use **zwei:command-store** or **zwei:set-comtab** to do so. You will want to put it into a command table which already exists, or one you create (for special-purpose editors you might write: See the section "Making Standalone Editor Windows" in *Programming the User Interface, Volume B*.) The standard comtabs you might be interested in modifying include:

zwei:*standard-comtab*
   all of the standard Zwei commands.

zwei:*zmacs-comtab*
   the Zmacs commands, like those having to do with buffers and files.

zwei:*zmail-comtab*
   the Zmail commands (you should define these with **zwei:define-zmail-top-level-command** instead of **zwei:defcom**, which takes different options, chosen from **zwei:no-sequence-ok**, **zwei:no-msg-ok**, **zwei:must-have-msg**, **zwei:numeric-arg-ok** or **zwei:no-arg**).

To see how these are set up, try looking at **zwei:initialize-standard-comtabs** and **zwei:initialize-zmacs-comtabs**.

12.2.3 Reading From the Mini-buffer

Another set of tools often used in writing editor commands are the functions for reading from the mini-buffer. There are many – and most of them are obsolete, special cases which were designed before the invention of **accept**.

The proper tool to use in most cases is **zwei:typein-line-accept**, which works just like **accept** with a few exceptions. Most notable is that it takes two extra keyword arguments which allow you to specify what the user has "already typed in" before it gets called. These keywords are **:initial-input** and **:initial-position**.

A few others remain for the intrepid:

zwei:read-function-spec reads a function specification from the mini-buffer. Clicking on function presentations, including those in the editor buffer, will work. It takes a prompt string as an argument, and optionally a default and a flag (**must-be-function**) which says what kind of function is acceptable: **nil** means anything which might be a function name, **t** means it must refer to a defined function, and **zwei:lambda-ok** means it is either a defined function or a lambda expression.

zwei:typein-line-history-readline reads a line from the user, saving the result in an input history.

zwei:typein-line-history-read reads a lisp form from the user, saving the result in an input history.

zwei:read-buffer-name and **zwei:read-defaulted-pathname** do what you'd expect.

12.2.4 A Real Example

Here's something taken out of the editor code, the definition for m-X Insert Date.[9]

```
(DEFCOM COM-INSERT-DATE
        "Print the curent date into the buffer.
Calls TIME:PRINT-CURRENT-TIME, or if given an
argument TIME:PRINT-CURRENT-DATE"
        ()
  (LET ((STREAM (OPEN-INTERVAL-STREAM (POINT))))
    (FUNCALL (IF *NUMERIC-ARG-P*
                 #'TIME:PRINT-CURRENT-DATE
                 #'TIME:PRINT-CURRENT-TIME)
             STREAM)
    (MOVE-MARK (POINT))
    (MOVE-POINT (SEND STREAM ':READ-BP)))
  DIS-TEXT)
```

This command behaves differently depending on whether you have provided a numeric argument or not. It uses **zwei:open-interval-stream**, an internal version of **zwei:open-editor-stream**.

12.3 Learning More About the Editor

There is a very powerful tool for learning how the editor works. It's invoked with m-X Edit Zmacs Command. Use it. There's only one way to really learn to write editor extensions, and that's to read others.

---

[9]I was originally going to use m-X Evaluate Into Buffer for this example, but it was too long to fit adequately onto the page. If you want to read that function, type m-X Edit Zmacs Command and then m-X Evaluate Into Buffer.

12.4 Fun and Games

From *The Hacker's Dictionary*, Guy L. Steele, Jr., *et al*:

FOO (*foo*)

1. *interjection.* Term of disgust. For greater emphasis, one says **MOBY FOO** (see **MOBY**).

2. *noun.* The first metasyntactic variable. When you have to invent an arbitrary temporary name for something for the sake of exposition, **FOO** is usually used. If you need a second one, **BAR** or **BAZ** is usually used; there is a slight preference at MIT for bar and at Stanford for baz. (It was probably at Stanford that bar was corrupted to baz. Clearly, bar was the original, for the concatenation **FOOBAR** is widely used also, and this in turn can be traced to the obscene acronym "FUBAR" that arose in the armed forces during World War II.)

   Words such as "foo" are called "metasyntactic variables" because, just as a mathematical variable stands for some number, so "foo" always stands for the real name of the thing under discussion. A hacker avoids using "foo" as the real name of anything. Indeed, a standard convention is that any file with "foo" in its name is temporary and can be deleted on sight.

BAR (*bar*)

The second metasyntactic variable, after **FOO**. If a hacker needs to invent exactly two names for things, he almost always picks the names "foo" and "bar."

12.5 Problem Set

Questions

1. Write **com-comment-out-lines-in-region** and **com-uncomment-lines-in-region** to insert (and remove) semi-colons at the beginning of each line in the region. (Both of these already exist as parts of **zwei:com-comment-out-region**, but that version includes lots of hair for handling messy cases. Write something simple.)

2. Write a macro which can be used either inside or outside the editor, which redirects all output during the execution of its body to a newly-created editor buffer

3. Write the command **com-change-expression-style-to-bold**, which does what the keyboard macro defined above does, namely changes the character style in the expression to the left of the cursor to boldface.

4. [For extra credit: write **com-change-expression-style-to-bolder**, which makes, for example, **(nil nil nil)** into **(nil :bold nil)**, and **(nil :italic nil)** into **(nil :bold-italic nil)**].

Solutions

1. Comment out lines:

```
(defcom com-comment-out-lines-in-region
        "Comments out each line in the region."
        ()
  (region-lines (start end)
    (loop for line = start then (line-next line)
          until (eq line end)
          do (insert (create-bp line 0) #\;)))
  dis-text)
```

Uncomment lines:

```
(defcom com-uncomment-lines-in-region
        "Removes semicolons from beginning of each line
in region."
        ()
  (region-lines (start end)
    (loop for line = start then (line-next line)
          until (eq line end)
          when (and (> (line-length line) 0)
                    (char-equal (aref line 0) #\;))
            do (let* ((end-idx (string-search-not-char
                                  #\; line 1))
                      (start-bp (create-bp line 0))
                      (end-bp (if end-idx
                                  (forward-char start-bp
                                                end-idx)
                                  (end-line start-bp))))
                 (delete-interval start-bp end-bp t))))
  dis-text)
```

Don't forget to add the commands to a comtab so you can
use them.

```
(set-comtab
  *zmacs-comtab*
  '(#\super-\; com-comment-out-lines-in-region
    #\hyper-\; com-uncomment-lines-in-region)
  (make-command-alist
    '(com-comment-out-lines-in-region
      com-uncomment-lines-in-region)))
```

2. Anything sent to **zl:standard-output**[10] during execution of *body* will be inserted into a buffer named *buffer-name*. There will also be messages inserted before and after *body* is executed.

```
(defmacro with-output-to-editor-buffer ((buffer-name)
                                            &body body)
  '(with-editor-stream (standard-output
                        :buffer-name ,buffer-name)
    (format t "~2%;;; Diverting to buffer (~\datime\)~2%")
    (multiple-value-prog1
      (progn ,@body)
      (format t "~2%;;; End of diversion (~\datime\)"))))
```

Note that this macro returns the values of the body. This is good practice whenever you write macros whose names start with **with-**.

3. Examine how c-X c-J is implemented to see how this works.

---

[10]remember, **zwei:** is a *Zetalisp* package. The Common Lisp variable ***standard-output*** is changed whenever you bind or otherwise modify **zl:standard-output**.

```
(defcom com-change-expression-style-to-bold
                "Changes one SEXP's style to (nil :bold nil).
Negative arguments are forward, positive backwards,
unlike all other commands."
                ()
  (let ((bp (forward-sexp (point) (- *numeric-arg*))))
    (change-style-interval (point) bp nil
        (si:style-index '(nil :bold nil) t))))
```

4. I have not solved the extra credit problem in the general case. Here is a special case version of it for my particular application, which was to convert comments to italics.

```
(defvar *default-comment-char-style-alist*
        '(((nil nil nil) (nil :italic nil))
          ((nil :bold nil) (nil :bold-italic nil))))

(defcom com-change-style-to-comment
        "Changes the style of the comment on this line
to italics.  A kludge."
  ()
  (let*
    ((line (bp-line (point)))
     (start (find-comment-start line t))
     (comment-start (and start (create-bp line start)))
     (comment-end (end-line (point))))
    (if (not comment-start)
        (barf "No comment on this line.")
        (undo-save comment-start comment-end t
                "change character style")
        (loop for (from-style to-style) in
                *default-comment-char-style-alist*
            as from-style-index =
                (si:style-index from-style t)
            and to-style-index =
```

```
                (si:style-index to-style t)
            do
        (change-one-style-interval-internal
          comment-start comment-end t
          from-style-index to-style-index)))
   dis-text))


(defun change-one-style-interval-internal
       (start-bp end-bp in-order-p from-style to-style)
  (get-interval start-bp end-bp in-order-p)
  (mung-bp-interval start-bp)
  (do
    ((line (bp-line start-bp) (line-next-in-buffer line))
     (limit-line (bp-line end-bp))
     (start-index (bp-index start-bp) 0)
     (last-line-p))
    (nil)              ; Don't return here, only at bottom.
    (setq last-line-p (eq line limit-line))
    (or (zerop to-style) (string-fat-p line)
        (setq line (set-line-array-type
                     line 'art-fat-string)))
    (setq line (mung-line line))
    (let ((line line))
      (declare (sys:array-register line))
      (do ((index start-index (1+ index))
           (limit-index (if last-line-p
                            (bp-index end-bp)
                            (line-length line))))
          ((≥ index limit-index))
        (let ((ch (aref line index)))
          (when (= (si:char-style-index ch) from-style)
            (setf (si:char-style-index ch) to-style)
            (setf (aref line index) ch)))))
    (and last-line-p (return nil))))
```

Ideally, you'd want to have an internal function to which you would pass a function which would return a new style given the current style of the character; this would prevent the mess which might happen if you happen to convert into a style which was about to be converted out of again.

13. A Quick Look At the Network

Although it is common to refer to a Lisp Machine's connections to the rest of the world as "the network," as if the machine were connected via a single mechanism to a unified system of linkages, such is not the case. There are several means of communication, operating via several different hardware and software protocols. And there is considerable overlap, with different software protocols operating simultaneously over the same hardware. It's not very complicated, but it's easy to become highly confused if the basic issues are not kept clear.

13.1 The Gee-whiz Look

This section might be better titled "How the network can be completely invisible to the user." You've probably been using the network for quite some time, usually without thinking about it. When you first log in, your machine connects to the namespace server to obtain information about you from the namespace database. Then your machine contacts your file server to load your init file. You start working on a program by editing a file containing the source; it comes from (perhaps) a different file server. And you send mail to your coworkers about problems you may be having with their software (or answer their complaints about yours!).

All of these actions require the use of "the network," unless you have a single machine not connected to any other. The point is, you can't tell whether the network is in use or not, in general, because programs such as the editor, the tape dumper, the mailer, and so forth, are written in such a way that the network level is immaterial; by the time the editor sees an interface to the network, it's a file stream, and behaves the same regardless of whether the file is on the local machine, the file server next door, or a machine across the country.

13.1.1 What Is a Network?

There are many ways to think about computer networks. An informal definition of a network might say something to the effect that it's a way for computers to talk to each other. However, computers are not like human beings: a pair of people can get together and have independent conversations. On the other hand, computers need to be told what to do, very explictly. In a computer network, one computer, or, to use the "network-ese" word for it, one *host*, tells another exactly what to do.

The model used in the Symbolics network system is based on *services*. A computer requests a service from another, which provides it. The first computer, called the *user* host, says very explictly what it wants the other computer, the *server* host, what to do. A service can be pretty much anything a remote computer might be expected to do for you, anything from storing your files on its disk to operating a robot for you.[1] Services can also be used automatically by programs, such as mail delivery programs which store mail destined for remote sites.

Since we're talking about a Lisp Machine, it's not surprising that the service model behaves very much like Lisp functions. Invoking a service via a network is very much like invoking a function: it might return values, perform side effects, and so forth.

You might imagine a number of ways to implement a given service, but there are two basic dimensions which might characterize a given implementation. In Symbolics' terminology, these are called *Medium* and *Protocol*. A medium, conceptually, is a combination of hardware and software which permits you to communicate at all. A network protocol is a software agreement as to what you actually say on the communication medium in order to make your needs known.

Consider a service that, say, the host named "Paul-Revere" might perform for the host "Old-North-Tower:" the *Alert Farmers of Invasion* service. This service might return no interesting values, but would have the side effect, say, of waking all the users on hosts at the "Middlesex" site.

The traditional implementation of the *Alert Farmers* service can be characterized as follows (figure 8):

- Medium: Lights on the top of Old North Church

- Protocol: One if by land, two if by sea.

---

[1] Years ago, at the MIT AI lab, there was a host which would respond to "fetch an elevator" service requests by bringing an elevator to your floor. You could invoke this service just before leaving for the night, and know there would be an elevator when you got out of the lab.

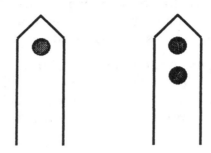

Figure 8.   Standard medium and protocol for *Alert Farmers*.

Now, each of these dimensions is independent of the other. For
example, you could use the traditional protocol, "one if by land
..." but change the medium to flags instead of lights. Or, you
could preserve the medium, namely lights on top of the tower,
but change the protocol to "yellow over red if by land, red over
yellow if by sea." You can even change both, and use both the
flag medium and the color protocol.[2]

13.1.2 Levels of Abstraction

When network hackers discuss *levels of abstraction*, they merely
mean that each layer of a network system is built on top of
others. It is often helpful to implement large systems in terms
of layers of modularity, and networks are particularly sensitive
to this.

---

[2]Both of these "media" use the same network hardware, namely the church tower.
You can imagine a medium which involves lights, but not the same hardware, such as
a rowboat across the river carrying lanterns. Similarly, certain network media can be
implemented on top of different lower-level hardware and software substrates.

| | **Medium** | |
| **Protocol** | **Lights** | **Flags** |

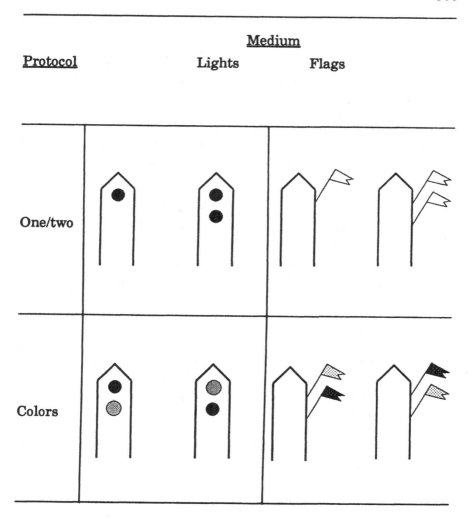

One/two		
Colors		

Figure 9. All possible implementations of *Alert Farmers*.

Here are some of the layers of network software, in increasing modularity. Each of these uses the previous layer to get its job done.

- The layers you don't need to know about: how many wires are in the cable, what the electrical characteristics of the connection are, etc. These are sometimes called the "Physical" layers.

- The "Network" layer: responsible for attempting to get data from here to there. Concerned with routing, addressing, etc.

- The "Transport" layer: responsible for getting data from there to there which is more-or-less correct. Depending on the transport layer's definition, this might mean that the bits are defined to arrive in the correct order, or all be correct, or whatever. For applications where you don't care that all the bits get there, like transmission of speech, you might use a transport layer which promises that, say, over 85% of the bits arrive.

- The "Session" layer: responsible for maintaining connections between cooperating processes on different hosts. Can create the cooperating process on the server host if it needs to.[3]

- The "Application" layer: responsible for converting between application-specific data and a representation which can be packaged and shipped via the transport layer, on connections provided by the session layer.[4]

---

[3]The layers up to this point correspond approximately to the "medium" portion of obtaining service.

[4]This layer corresponds to the "protocol" part. There may be some higher-level "medium" aspects here as well. For example, consider the :byte-stream-with-mark network medium. See the section "BYTE-STREAM-WITH-MARK Network Medium" in *Networks*.

• Application programs: use interfaces provided by the application layer to communicate intent, desires, etc.

In general, implementing earlier (lower) levels requires knowing more, and being able to hack closer to the hardware, etc.

The Symbolics documentation is very good about describing how to write application programs which use the application layer. It's also pretty good at describing how to write new application layers, although not quite as good. It's pretty hard to write new network, transport and session layers from the documentation, but at least one Symbolics customer did it a couple of years ago, and the documentation is considerably better now.

13.2 The Generic Network System

The *Generic Network System* (GNS) is concerned with the part of the network system up to the the application layer. It provides an interface between the application programs and the various session and transport layers.

Just as the messenger who runs up to Old North Church and tells the "host" there that she wants to invoke the *Alert Farmers* service doesn't really care what the medium and protocol are, so the application program which invokes the service doesn't really care either.[5]

The Generic Network System provides two classes of interfaces:

1. Finding a path to a service: In order to invoke a service, you must be able to discover the protocols and media

---

[5]The GNS provides hooks which permit programmers who really *do* care to invoke specific implementations of various services.

which implement it, and choose among them based on how desirable the implementations are.

2. Invoking a given *service access path*, as found by the previous step.

The function **net:invoke-service-on-host** combines these two steps into a single function call, which is usually what you want. The only times you would want to worry about access paths is when you

- are doing something fancy with multiple hosts,
- don't care which host provides the service, or
- are going to do something fancy depending on which protocol gets used.

13.2.1 How Does Path-finding Work?

Finding a path to a service means figuring out exactly what must be done in order to invoke a given service on a given host. There are functions which find a path to the given service on a specific host, on any host, or find a way to invoke the service via a "broadcast" mechanism.

In order to do its job, the path-finding function must consider three separate sets of information.

1. Does the remote host implement the service? If so, what protocol must one use, and on which media will that protocol be useable?

2. How is the network configured? In other words, what possible network connections exist between the local and remote hosts which will support one or more of the desired media?

3. What protocols and media does the local host know how to use to invoke the given service?

The first two sets of data come from the *namespace database*, the place where network configuration is kept. The third is stored in the local host's Lisp world, and is dependent on which software is loaded.

The result is a (possibly empty) list of service access path objects, sorted in order of desirability. Desirability is measured in terms of how fast the network connection is, how powerful the protocol is, and so forth.

13.2.2 How Does Service Invocation Work?

Service invocation is really the easy part of the process, once you know how to get there:

1. It calls upon the services of the transport layer indicated in the medium description to form a connection, and

2. It invokes the function which implements the protocol, passing it a medium-specific handle to the network connection (this handle is often a stream).

13.2.3 Other GNS Functions

In order to invoke a service on a host, you must first have a *host object* which represents that host. If you want to convert a string into a host object, you call **net:parse-host**:

```
(net:parse-host "CD")
```

Usually, **net:parse-host** and **net:invoke-service-on-host** are all you really need. However, there are plenty of other hooks into the Generic Network system:

- **net:find-paths-to-service-on-host** – returns a list of *service access paths*, sorted in order of desirability.

- **net:find-path-to-service-on-host** – returns the most desirable service access path.

- **net:find-paths-to-service** – returns a list of service access paths to all hosts which support the requested service.

- **net:find-paths-to-service-using-broadcast** – returns a list of service access paths which "broadcast" your request to all hosts on your network. You would use this when you're trying to get service from any host, and don't mind disturbing potentially every host on the network to get it. Only services which are inexpensive to supply, such as time of day, are invoked using the broadcast mechanism.

- **net:find-paths-to-protocol-on-host** – returns service access paths which implement a service using a particular protocol.

- **net:invoke-service-access-path** – does what it says.

- **net:invoke-multiple-services** – allows you to invoke a list of services at once. You might use this to find out who's logged into all the Lisp Machines at your site at once. To see how this is used, read the function **neti:*scan-lispms**.

13.3 The Namespace System

The namespace system is a distributed[6] database system which was designed primarily for the Generic Network System. For this purpose, it contains two items of interest about hosts and network topology:

1. *Service* attributes of hosts: a list of triples, consisting of the service, the medium on which it is offered, and the protocol by which it may be used.

2. *Network addresses* of hosts: a list of pairs, consisting of the network to which the host is connected, and its address.

As long as a fairly robust, distributed database system has been implemented, we may as well use it to store other information. Thus, in addition to host objects, the namespace database contains information about a variety of other classes of objects. Here are some other things kept in the namespace database:

- Other attributes of *hosts*: extra names, "pretty" names (the mixed-case name), its operating system, its console location, various peripherals attached to it, printers for which it is the spooler, etc.

- *Site* attributes: its name, security information, mail characteristics, the local timezone, "pretty name," etc.

- *User* attributes: a login name, a full name, home and work addresses and phone numbers, that person's birthday, etc. This is data commonly given out for "Finger" network service.

---

[6]"Distributed," in this context, means that it is distributed over a number of different hosts. Each host has some local information about the database, and knows which host(s) to ask for more data.

- *Printer* attributes: what kind of printer it is, its location, the host to which it is attached (and via what hardware), and so forth.

- *Network* attributes: its name, what type of network it is, its special topology requirements, etc.

- *Namespace* attributes: who are its servers, what other namespaces it includes by reference, etc.

13.4 Examples of the Use of the Generic Network System

Here are some sample application programs which use the network to get their work done. In many cases, the "application program" at the level that I will describe it has already been written, usually more comprehensively than I will sketch out. I will point out places in the system source you might look for further information. Reference documentation for using network services is in volume 9: See the section "How a Network Service is Performed" in *Networks*.

13.4.1 Time of Day

To get the time of day from a specific host:

```
(net:invoke-service-on-host :time host)
```

This service has no side-effects, and returns the time of day, in universal time (seconds since January 1, 1900).

A function which does this in a much fancier fashion is **net:get-time-from-network**. It *broadcasts* its request, which is a way of asking every host on a given network (or part of a network) to respond simultaneously. Broadcast services usually only look at the first result returned, which is why they should only be used for very inexpensive or very infrequent applications.

13.4.2 Who's Logged In

To find out who's logged in to a specific host, in as much detail as that host can supply:

```
(net:invoke-service-on-host :show-users host
                              :whois t)
```

This service returns no interesting values, but as a side-effect it prints the logged-in users from the foreign machine on the stream **standard-output**. There are two optional keyword arguments you can supply to this service:

- **:whois** — Return all possible information, such as addresses and phone number, birthday, and so forth. If the host is a Lisp Machine, this data will come from whatever the user has filled in in the namespace database.

- **:user** — Return information about the specific user. The default is to return information on every currently logged-in user.

For a system-supplied function which uses this service, see the source for **net:finger**.

13.4.3 Mail Delivery

This version of mail delivery is considerably simpler than the system-supplied one. The system-supplied macro **mailer:with-mailer** takes a service access path as an argument, while the one I'm about to define takes a host. The body of the macro form should be identical in both cases.

As mentioned earlier, the *invocation* of a mail service (either **:store-and-forward-mail**, which tries to deliver the mail no matter where it goes, or **:mail-to-user**, which doesn't guarantee that it will try, although it might) returns a mailer object and has no direct side effects. However, sending messages to the mailer object is how you send mail, which usually entails side effects.

The messages you can send to a mailer object include:

- **:start-message** – Takes the sender as an argument.

- **:verify-recipient** – Takes a recipient as an argument. [Both sender and recipient are examples of a somewhat complicated data structure, returned by such functions as **zwei:parse-addresses** and **zwei:parse-one-address**].

- **:receive-message** – Takes a "trace line" (which goes at the front of the mail, usually a "Received:" line or **nil**), a header stream, and an optional body stream (if not supplied, the first stream is supposed to have all the characters which go in the entire message).

- **:finish-message** – Tells the mailer host to actually deliver the mail.

    ;;; First, a macro which wraps up the important stuff.

```
(defmacro with-hack-mailer ((mailer host) &body body)
  '(let ((,mailer nil))
     (unwind-protect
         (progn (setq ,mailer
                      (net:invoke-service-on-host
                       :store-and-forward-mail
                       (net:parse-host ,host)))
                ,@body)
       (when ,mailer (close ,mailer))))))

(defun hack-mail (host recips msg-string)
  (with-hack-mailer (mailer host)
    (send mailer :start-message
          (zwei:parse-one-address
           (format nil "~{~A@~A~}" ; "foo@bar"
                   (send si:*user* :mail-address))))
    (loop for recip in (zwei:parse-addresses recips)
          do (send mailer :verify-recipient recip))
    (with-input-from-string (msg-stream msg-string)
      (send mailer :receive-message nil msg-stream))
    (send mailer :finish-message)))
```

As I indicated earlier, a much more complete macro is **mailer:with-mailer**. Instead, of a host, it takes a service access path, so the function **hack-mail** above might have been written as follows:

```
(defun hack-mail (host recips msg-string)
  (mailer:with-mailer
    (mailer (net:find-path-to-service-on-host
             :store-and-forward-mail
             (net:parse-host host)))
    [...]))
```

13.5 Writing Your Own Network Software

Now, it's all well and good to use software which has already been written, but sometimes you want to make the machine do something new. In networks, this often involves designing and implementing a new protocol to perform that service.

What does this involve? In general, for a network protocol to be successful, you must have both ends agree on each of these points:

- Connection name: the two ends need to agree on how the initial connection is established. For TCP, for example, the two ends need to agree on a "well-known port." For Chaosnet, they must agree on the "contact name." In any case, the user end invokes the session layer with the appropriate name for the protocol it wishes to speak.

- Data representation: in order to transmit your data, you must encode it in some form which is amenable to transmission on your network medium. For example, if your medium transmits 8-bit bytes, you must convert integers, strings, and other Lisp objects into some representation which fits in integers between 0 and 255. At the receiving end, you must decode it into some useful representation. You can use the same format for transmitting data in both directions, or have the two ends use completely different formats.

- Command/response format: how will the user end tell the server end what to do? How will the server end reply that it has been done? How can it say that there was an error, and what happened instead of the desired result? Again, your commands and responses must be encoded into a form that your medium can transmit.

- Auxilliary connections: you might like to be able to interrupt transmission of long streams of data without having
  to worry about the integrity of the data. One way to do
  this is to reserve the original connection for sending commands and interrupts, and create auxilliary connections
  for the actual data. This is the way most file transfer
  protocols work, for example. The two ends need to agree
  on the "names" of these connections, or a mechanism for
  communicating the name from one end to the other.

- Which end speaks first: again, computers need to be told
  exactly what to do. If each end waits for the other to say
  something before sending their first data, your protocol
  isn't going to work. Similarly, you have to have an
  agreement on which end waits until the other is completely finished.

- Which end speaks last: a particularly nasty timing
  problem is found at least in the Chaos network medium.
  When both ends try to close a connection "cleanly," each
  of them waits for the other to acknowledge that all the
  data has been received before either of them will say any
  more. In order to get around this, one end must acknowledge all data, including the End-of-File marker, from the
  other. One way to do this is for one side to forcibly close
  the connection (*i.e.*, use the **:abort** keyword when closing
  the stream). A cleaner way is for one side to explicitly
  read the other's EOF packet, by using
  stream-copy-until-eof to copy from the stream to the
  "null stream."

If the protocol is intended to operate independently of human
intervention, its command data representations should be easy
for machines to parse. For example, consider the mail transfer

protocol SMTP[7]. This protocol is used by mail transmitting and receiving programs, and is mostly expected to run autonomously. In SMTP,

- all command names are four letters long, and are followed by data in a very constrained format;
- responses are all three-digit numbers, with human-readable strings attached to error response codes; and
- only a very small number of commands and responses are valid at any point in the dialog between the user and server programs.

In contrast, consider the remote login protocol named Telnet[8]. In this protocol, the user is typing most of the input, and is reading and (presumably) responding to the output. Thus, Telnet commands and responses, for the most part, are very unconstrained.

For a new protocol, you will need to write both a user end and a server end. If you are only planning to run, say, the user end on the Lisp Machine (your server might be some other kind of host), you only need to write the user end on the Lisp Machine. For my own work, I find that writing both ends on the Lisp Machine helps to clarify the issues, and makes it easier to debug your protocol. Writing the "foreign" user or server after that is easier, of course, because the algorithm is already debugged.

For complete documentation on writing user and server network software: See the section "Defining a New Network Service" in *Networks*.

---

[7]The Simple Mail Transfer Protocol, defined in ARPAnet RFC 821.

[8]ARPAnet RFC 854.

13.5.1 Writing Your Own User End

On the Lisp Machine, the way to write your own user end is with the macro **net:define-protocol**. See the special form **net:define-protocol** in *Networks*. Writing a protocol is usually very straightforward.

Most network protocols usually take place on a single network connection, usually a reliable byte-stream type of connection. Certain protocols will want to use a more sophisticated layer on top of the byte stream, for reasons of synchronization or to aid in conversion of Lisp objects into a representation which can be placed into a byte stream. A couple of these are already in place in Genera: **:byte-stream-with-mark** and **:token-list** media are implemented by formatting data carefully onto byte streams provided by TCP and Chaos, for example.

Usually the connection is used in both directions, often one at a time in a sort of "lock-step" fashion. That is, the user end sends a single command, and waits for a single response from the server. The server waits for a command, and sends its response to that command.

13.5.2 Writing Your Own Server End

Writing a server is the converse of writing a user-end. It is also very straightforward, and is done with the macro **net:define-server**. See the special form **net:define-server** in *Networks*. Once again, both ends of the connection must agree on all the details of the protocol.

One undocumented keyword you can supply to **net:define-server** is **:flavor**. If you do this, then you don't have to define the server as the body of **net:define-server**. Instead, an instance of that flavor is created, and a process is created which sends that instance the message **:server-top-level**. Your flavor should be based on the flavor **net:byte-stream-server** or one of its components. For servers which will stay around for a long time, such as file servers, this is often a good way to associate state variables with the server, by using the instance variables of the server instance. See, for example, the way the **:namespace** server is implemented (type meta-. :namespace followed by meta-0 meta-. until you find the right definition of **:namespace**).

13.5.3 Sample User and Server Definition

Here is an implementation of a user-information service, a sort of stripped-down "WHOIS" facility.

It might be invoked by a program doing the following on the user host:

```
(defun who-is-using (host)
  (setq host (net:parse-host host))
  ;; Convert it into host object
  (let ((return-plist (net:invoke-service-on-host
                        :user-info host
                        :name t :work-phone t)))
    (format t "~%~A is using ~A; work phone is ~A"
            (getf return-plist :name)
            (send host :pretty-name)
            (getf return-plist :work-phone))))
```

The protocol is quite simple, and can be implemented on any byte-stream medium. Each interaction between the user and the server consists of a single line of text (a "command") from the user, which must be one of "NAME", "ADDRESS", "HOME PHONE" or "WORK PHONE". The server replies with a single line, which is either the character #\+ and a response, or the character #\- and an error message.

Note that the entire interaction takes place in the Lisp Machine character set, and lines are delimited with Lisp Machine #\return characters. It is possible to define both a user and a server which communicates in ASCII characters instead, in which case the end-of-line character sequence would be CR-LF (or just CR). [This latter is completely hidden from the writer of the protocol, once s/he says :ascii-translation, the matter is completely taken care of.]

13.5.3.1 User-end Protocol Definition

The user end of the protocol is defined with net:define-protocol. This definition must be loaded into the user host's lisp environment.

```
(net:define-protocol :user-info-protocol
                     (:user-info :byte-stream)
  (:invoke-with-stream-and-close
    (stream &key name address home-phone work-phone
            &aux result)
    (flet ((send-command (command)
             (format stream "~A~%" command)
             (send stream :force-output)
             (let ((response (send stream :line-in)))
               (when (< 0 (string-length response))
                 (selector (aref response 0) char-equal
                   (#\+ (substring response 1))
                   (#\- (error "Command failed: ~A: ~A"
                               command
                               (substring response 1)))
                   (otherwise
                     (error "Invalid response: ~A"
                            response)))))))
      (when name
        (push :name result)
        (push (send-command "NAME") result))
      (when address
        (push :address result)
        (push (send-command "ADDRESS") result))
      (when home-phone
        (push :home-phone result)
        (push (send-command "HOME PHONE") result))
      (when work-phone
        (push :work-phone result)
        (push (send-command "WORK PHONE") result))
      (send-command "BYE")
      (values (nreverse result)))))
```

flet is used to introduce a local function without having a
globally-named one; it is similar to **let**.   In this case, the

send-command function is used as the central control point to ensure that the protocol is followed correctly.

13.5.3.2 Server-end Protocol Definition

The server end of a protocol is defined with **net:define-server**. Note that the server definition doesn't say what service it's supplying. A server for a particular protocol might actually implement several services. For example, the SMTP server implements **:mail-to-user** and **:store-and-forward-mail**, which are not very different. It also implements **:expand-mail-recipient**, which is a completely different service. The **net:define-protocol** form, of course, needs to know which service it's performing.

This definition is loaded into the server host's environment.

```
(net:define-server :user-info-protocol
    (:medium :byte-stream :stream stream)
    (labels ((command-response (ok? result)
                (format stream "~:[-~;+~]~A~%" ok? result)
                (send stream :force-output))
             (answer (info)
                (command-response t (send si:*user* info))))
      (loop named server
            for command = (send stream :line-in)
            do (selector command string-equal
                ("NAME" (answer :personal-name))
                ("ADDRESS" (answer :work-address))
                ("HOME PHONE" (answer :home-phone))
                ("WORK PHONE" (answer :work-phone))
                ("BYE"
                 (command-response t "See you later")
                 ;; Deal with EOF synchronization.
                 (stream-copy-until-eof
                             stream #'sys:null-stream)
                 (return-from server (values)))
                (otherwise
                  (command-response
                    nil
                    (format nil "Invalid command: ~A"
                            command)))))))
```

13.5.3.3 Contact Names

The other part of the agreement between the user and server ends is how they begin to talk to each other. Each kind of transport medium has its own characteristic way to do this.

At least one of these must be loaded into both the user and server hosts' environments.

```
;;; Here is how you hook up a protocol and a Chaos contact
;;; name, TCP protocol "well-known port", or DNA contact id.

(chaos:add-contact-name-for-protocol :user-info-protocol)

(tcp:add-tcp-port-for-protocol :user-info-protocol 123)

(dna:add-dna-contact-id-for-protocol :user-info-protocol
                                       "USERINFOPROTOCOL")
```

13.5.3.4 Updating the Namespace

So far, the user host knows how to invoke the service, and the server host knows how to respond correctly to the protocol. Also, they both agree on how the initial contact is to be supplied.

Now, how does the user host find the path to service? The answer is, you must update the namespace database to tell it.

Presumably, the two hosts are already recorded in the namespace, and even have network addresses which allow them to communicate. All that remains is telling the user's host how the server provides the service. To do that, you must add a *service* entry to the server host's namespace entry. For the **:user-info** service we defined above, you would need to add one or more of the following service entries to the server host:

```
USER-INFO  CHAOS  USER-INFO-PROTOCOL
USER-INFO  TCP    USER-INFO-PROTOCOL
USER-INFO  DNA    USER-INFO-PROTOCOL
```

13.5.3.5 Debugging the Protocol

When I first wrote this example, it had two bugs:

1. I left off the "~%" in the **format** call in **send-command**, and also in **command-response**. Since the other end was

expecting to read lines using **:line-in** messages, this was a violation of the protocol.

2. Each end was waiting for the other to synchronize the stream closing. In Chaos protocol, what happens is that each sends an EOF packet, and waits for the other to acknowledge it. Since neither is talking any more after it sends the EOF, no packet with an acknowledgement ever gets sent. I am not sure whether other network media have similar timing problems, but I have found that it always makes sense for one end or the other to read the final EOF packet, using **stream-copy-until-eof** to the null stream.

Appendix A
Basic Zmacs Commands

While it is true that there are a great many Zmacs commands, and trying to learn them all would be a close to hopeless task, it is also true that the situation is really much less difficult than it might at first appear. For one thing, you can edit files quite effectively with a relatively small subset of the Zmacs commands. (This should not be taken to imply that the remaining commands are superfluous; the "quite effective" editing you can do without them is transformed into astonishingly effective editing with them.) Another saving grace is that there is some pattern to the pairing of keystrokes with editing functions. For instance, *control* characters often act on single letters or lines; *meta* characters on words, sentences, or paragraphs; and *control-meta* characters on lisp expressions. Thus c-F moves forward one character, m-F moves forward one word, and c-m-F moves forward one lisp expression. c-K means "kill" (delete) to the end of the line, m-K means kill to the end of the sentence, and c-m-K means kill to the end of the current lisp expression. So the amount of memorizing you have to do to start editing is really not very great.

I can't overemphasize the utility of the Help facility in Zmacs. It can be a real lifesaver, both when you don't know what commands there are to do something, and when you've forgotten how to invoke a command you know about. So don't limit yourself to the commands listed below. Consider the list a crutch, to help get you started, but try to leave it behind as soon as possible.

With Genera 7.0, Symbolics has introduced *Reference Cards*, a little spiral-bound notebook full of "cheat sheets." What follows is a distillation of the Zmacs section, with a few commands added (I've underlined those).

Many of these commands also work in the Lisp Listener, and are also useful for copying things from the editor to the typeout window or minibuffer. The underlined mouse commands are especially useful in this respect.

Movement Commands

c-F	Move forward one character
c-B	Move backward one character
c-N	Move down one line ("next")
c-P	Move up one line ("previous")
c-A	Move to the beginning of the line
c-E	Move to the end of the line
[mouse left]	Move to mouse position
m-F	Move forward one word
m-B	Move backward one word
m-A	Move to the beginning of the sentence
m-E	Move to the end of the sentence
m-[	Move to the beginning of the paragraph
m-]	Move to the end of the paragraph
m-<	Move to the beginning of the buffer
m->	Move to the end of the buffer

c-m-F	Move forward one lisp expression
c-m-B	Move backward one lisp expression
c-m-A, c-m-[	Move to the beginning of the current definition
c-m-E, c-m-]	Move to the end of the current definition

Deletion Commands

c-D	Delete forward one character
Rubout	Delete backward one character
Clear Input	Delete to the beginning of the line
c-K	Delete to the end of the line
m-D	Delete forward one word
m-Rubout	Delete backward one word
m-K	Delete forward one sentence
c-m-K	Delete forward one lisp expression
c-m-Rubout	Delete backward one lisp expression
c-Y	Restore ("yank") text deleted with any of the above, except c-D and Rubout
m-Y	Immediately following a c-Y or another m-Y, replace the yanked text with the previous element of the kill history

Region Commands

c-Space	Set the mark at the current position, and turn on the "region." Subsequent movement commands will define the region to be the area between the mark and the new position.
c-W	Delete the region, putting it on the kill history
m-W	Put the region on the kill history without deleting it
[mouse middle]	Mark (make into the region) the object the mouse is pointing at

[mouse left] hold
> Mark the area dragged over (between button press and button release)

control-[mouse middle]
> Copy the object the mouse is pointing at to the cursor. This is useful for copying whole forms; point at the open or close parenthesis and click c-[mouse-middle].

File Commands

c-X c-F
> Read ("find") a file into its own buffer, creating an empty buffer if the file doesn't exist

c-X c-S
> Write ("save") a buffer back to its file

c-X c-W
> Write a buffer to any file, specified by name

M-x Compile File
> Compile a file, *i.e.*, write a binary version of the file. This has no effect on the current Lisp environment. (Compare with M-x Compile Buffer.)

Buffer Commands

c-m-L
> Switch to the previous ("last") buffer; with a numeric argument you can get to other previous buffers.

c-X B
> Switch to a buffer specified by name

c-X c-B
> Display a list of the buffers

Lisp Commands

c-sh-E
> Evaluate (call the Lisp interpreter on) the region, if it's active, or the current definition if it's not

c-sh-C
> Compile the region or current definition into the Lisp environment

M-x Evaluate Buffer

Evaluate the entire buffer (usually you should just compile it)

M-x Compile Buffer

Compile the entire buffer into the environment. This has no effect on the file system. (Compare with M-x Compile File.)

End, s-E

Murray Hill Standard Utilities only. Evaluate the region or current definition and insert the result into the buffer.

Miscellaneous

Suspend
Enter the typeout window. (Resume returns.)

m-.
Find the definition of a given function

c-X D
Directory edit – shows a directory listing and enables manipulation of the files in it

Help A
Apropos – list all commands containing a given substring

Help C
Describe the command associated with a given keystroke

Help D
Describe a command specified by name

Index